I0476846

Cover: This photograph from 2007 shows a century old whaling boat that became perched along a segment of rapidly eroding shoreline on the Beaufort Sea coast. Two months after this photograph was taken the boat fell into the ocean and was washed away. (Photograph by Benjamin Jones, U.S. Geological Survey.)

The United States National Climate Assessment— Alaska Technical Regional Report

Edited by Carl J. Markon, U.S. Geological Survey; Sarah F. Trainor, University of Alaska Fairbanks–Alaska Center for Climate Assessment and Policy; and F. Stuart Chapin, III, University of Alaska Fairbanks

Circular 1379

U.S. Department of the Interior
U.S. Geological Survey

U.S. Department of the Interior
KEN SALAZAR, Secretary

U.S. Geological Survey
Marcia K. McNutt, Director

U.S. Geological Survey, Reston, Virginia: 2012

For more information on the USGS—the Federal source for science about the Earth, its natural and living resources, natural hazards, and the environment, visit http://www.usgs.gov or call 1–888–ASK–USGS.

For an overview of USGS information products, including maps, imagery, and publications, visit http://www.usgs.gov/pubprod

To order this and other USGS information products, visit http://store.usgs.gov

Suggested citation:
Markon, C.J., Trainor, S.F., and Chapin, F.S., III, eds., 2012, The United States National Climate Assessment— Alaska Technical Regional Report: U.S. Geological Survey Circular 1379, 148 p.

Preface

In 1990, Congress passed Public Law 101-606 (1990), which established the U.S. Global Change Research Program (USGCRP). The purpose of the USGCRP is to provide information that increases the understanding of the cumulative effects of human activities and natural processes on the environment and their response to global change. Section 106 of the Act identifies the requirement for a National Assessment to be delivered to the President of the United States and Congress not less frequently than every 4 years that:

- Integrates, evaluates, and interprets the findings of the Program and discusses the scientific uncertainties associated with such findings;

- Analyzes the effects of global change on the natural environment, agriculture, energy production and use, welfare, human social systems, and biological diversity; and

- Analyzes current trends in global change, both human-induced and natural, and projects major trends for the subsequent 25–100 years (Public Law 101-606, 1990).

The National Climate Assessment (NCA) serves as a status report on climate change science and impacts based on observations made across the country. It incorporates advances in the understanding of climate science into larger social, ecological, and policy systems and serves to integrate scientific information from multiple sources and to highlight key findings and significant knowledge gaps. The First National Assessment report was produced in 2000 and a second, Global Climate Change Impacts in the United States, was produced in 2009.

This document, The United States National Climate Assessment–Alaska Technical Regional Report, is one of eight regional reports that will provide input to the 2013 National Climate Assessment. It was produced through the leadership of the 2012 NCA Alaska Region Technical Report Writing Team (appendix A), but is the culmination of the efforts of many contributing authors who are recognized in appendix B. Discussions began in 2011 (fig. 1) and the process included four public outreach events. Two webinars[1] were hosted by the University of Alaska Fairbanks, Alaska Center for Climate Assessment and Policy (ACCAP–November 2011 and February 2012). Two workshops, one at the Alaska Tribal Conference on Environmental Management (October 2011), and one at the Alaska Forum on the Environment (February 2012) also were held. These outreach events allowed the public a venue to provide comment and input on report topics and content.

9/12/11
Kick off
Meeting

10/25/11
ACCAP
Webinar

1/15/12
Technical
Report
Contents

2/8/12
AFE Session
Selected
Findings

3/1/12
Final
Technical
Report to
NCA

10/1/11
Expressions
of Interest

11/8/11
ATCEM

1/15/12
Compilation
Draft Report
begins

2/13/12
ACCAP
Webinar

Figure 1. Timeline for producing the National Climate Assessment Alaska Technical Report. ACCAP, Alaska Center for Climate Assessment and Policy; AFE, Alaska Forum on the Environment; NCA, National Climate Assessment; ATCEM, Alaska Tribal Conference on Environmental Management.

---

[1]Recordings of these webinars are archived on the Alaska Center for Climate Assessment Policy (University of Alaska, Fairbanks, 2012b).

This page intentionally left blank.

Acknowledgments

This report was made possible by the combined efforts and leadership of the writing team (appendix A) and the dedication of the contributing authors (appendix B). Financial support was provided by the USGS Climate and Land Use Mission Area, the Office of the Regional Director, Alaska Area, and the kind support of all participating agencies. Special thanks go to the following individuals that provided peer review of the manuscript: Amanda Robinson, U.S. Fish and Wildlife Service; Robert Winfree, National Park Service; and Keith Boggs, Alaska Natural Heritage Program.

This page intentionally left blank.

Contents

Contents—Continued

Figures

Figures—Continued

Figures—Continued

Tables

Conversion Factors

Inch/Pound to Metric

Multiply	By	To obtain
Length		
inch (in.)	2.54	centimeter (cm)
inch (in.)	25.4	millimeter (mm)
foot (ft)	0.3048	meter (m)
mile (mi)	1.609	kilometer (km)
yard (yd)	0.9144	meter (m)
Area		
acre	4,047	square meter (m^2)
acre	0.4047	hectare (ha)
acre	0.4047	square hectometer (hm^2)
acre	0.004047	square kilometer (km^2)
square foot (ft^2)	929.0	square centimeter (cm^2)
square foot (ft^2)	0.09290	square meter (m^2)
square inch (in^2)	6.452	square centimeter (cm^2)
section (640 acres or 1 square mile)	259.0	square hectometer (hm^2)
square mile (mi^2)	259.0	hectare (ha)
square mile (mi^2)	2.590	square kilometer (km^2)
Volume		
cubic mile (mi^3)	4.168	cubic kilometer (km^3)
Mass		
ounce, avoirdupois (oz)	28.35	gram (g)
pound, avoirdupois (lb)	0.4536	kilogram (kg)
ton, short (2,000 lb)	0.9072	megagram (Mg)
ton, long (2,240 lb)	1.016	megagram (Mg)
ton per day (ton/d)	0.9072	metric ton per day
ton per day (ton/d)	0.9072	megagram per day (Mg/d)
ton per day per square mile [(ton/d)/mi^2]	0.3503	megagram per day per square kilometer [(Mg/d)/km^2]
ton per year (ton/yr)	0.9072	megagram per year (Mg/yr)
ton per year (ton/yr)	0.9072	metric ton per year

Conversion Factors—Continued

Metric to Inch/Pound

Multiply	By	To obtain
Length		
centimeter (cm)	0.3937	inch (in.)
millimeter (mm)	0.03937	inch (in.)
meter (m)	3.281	foot (ft)
kilometer (km)	0.6214	mile (mi)
meter (m)	1.094	yard (yd)
Area		
square meter (m^2)	0.0002471	acre
hectare (ha)	2.471	acre
square hectometer (hm^2)	2.471	acre
square kilometer (km^2)	247.1	acre
square centimeter (cm^2)	0.001076	square foot (ft^2)
square meter (m^2)	10.76	square foot (ft^2)
square centimeter (cm^2)	0.1550	square inch (ft^2)
square hectometer (hm^2)	0.003861	section (640 acres or 1 square mile)
hectare (ha)	0.003861	square mile (mi^2)
square kilometer (km^2)	0.3861	square mile (mi^2)
Volume		
cubic kilometer (km^3)	0.2399	cubic mile (mi^3)
Mass		
gram (g)	0.03527	ounce, avoirdupois (oz)
kilogram (kg)	2.205	pound avoirdupois (lb)
megagram (Mg)	1.102	ton, short (2,000 lb)
megagram (Mg)	0.9842	ton, long (2,240 lb)
metric ton per day	1.102	ton per day (ton/d)
megagram per day (Mg/d)	1.102	ton per day (ton/d)
megagram per day per square kilometer [(Mg/d)/km^2]	2.8547	ton per day per square mile [(ton/d)/mi^2]
megagram per year (Mg/yr)	1.102	ton per year (ton/yr)
metric ton per year	1.102	ton per year (ton/yr)

Temperature in degrees Celsius (°C) may be converted to degrees Fahrenheit (°F) as follows:

$$°F=(1.8×°C)+32.$$

Temperature in degrees Fahrenheit (°F) may be converted to degrees Celsius (°C) as follows:

$$°C=(°F-32)/1.8.$$

Acronyms and Abbreviations

ACCAP	Alaska Center for Climate Assessment and Policy
ACCER	Alaska Climate Change Executive Roundtable
AMSA	Arctic Marine Shipping Assessment
AOOS	Alaska Ocean Observing System
AYK	Arctic-Yukon-Kuskokwin
C4	Alaska Climate Change Coordinating Committee
$CaCO_3$	Calcium Carbonate
CDQ	Community Development Quota
CMIP3	Coupled Model Intercomparison Project Phase 3
CO_2	Carbon Dioxide
CSC	Climate Science Center
DoD	Department of Defense
DBO	Distributed Biological Observatory
DOC	Dissolved organic carbon
DOI	Department of the Interior
GOA	Gulf of Alaska
GRACE	Gravity and Climate Experiment
Gt/yr	Gigatons per year
IPCC	Intergovernmental Panel on Climate Change
LCC	Landscape Conservation Cooperative
LEO	Local Environmental Observer program
NCA	National Climate Assessment
NCCWSC	National Climate Change and Wildlife Science Center
NVDI	Normalized Difference Vegetation Index
NOAA	National Oceanic and Atmospheric Administration
NPR-A	National Petroleum Reserve – Alaska
NSR	Northern Sea Route
NSSI	North Slope Science Initiative
PDO	Pacific Decadal Oscillation
PRISM	Parameter-elevation Regression on Independent Slopes Model
QDR	Quadrennial Defense Review
RISA	Regional Integrated Sciences and Assessments
SERDP	Strategic Environmental Research and Development Program
SNAP	Scenarios Network for Alaska and Arctic Planning
SRES	Special Report on Emissions Scenarios
TEK	Traditional Ecological Knowledge
Tg/yr	Teragrams/year
UAA	University of Alaska Anchorage
UAF	University of Alaska Fairbanks
USGCRP	U.S. Global Change Research Program
USGS	U.S. Geological Survey
YOY	Young of year

The United States National Climate Assessment— Alaska Technical Regional Report

Edited by Carl J. Markon, U.S. Geological Survey; Sarah F. Trainor, University of Alaska Fairbanks–Alaska Center for Climate Assessment and Policy; and F. Stuart Chapin, III, University of Alaska Fairbanks

Executive Summary

The Alaskan landscape is changing, both in terms of effects of human activities as a consequence of increased population, social and economic development and their effects on the local and broad landscape; and those effects that accompany naturally occurring hazards such as volcanic eruptions, earthquakes, and tsunamis. Some of the most prevalent changes, however, are those resulting from a changing climate, with both near term and potential upcoming effects expected to continue into the future.

Alaska's average annual statewide temperatures have increased by nearly 4°F from 1949 to 2005, with significant spatial variability due to the large latitudinal and longitudinal expanse of the State. Increases in mean annual temperature have been greatest in the interior region, and smallest in the State's southwest coastal regions. In general, however, trends point toward increases in both minimum temperatures, and in fewer extreme cold days. Trends in precipitation are somewhat similar to those in temperature, but with more variability. On the whole, Alaska saw a 10-percent increase in precipitation from 1949 to 2005, with the greatest increases recorded in winter.

The National Climate Assessment has designated two well-established scenarios developed by the Intergovernmental Panel on Climate Change (Nakicenovic and others, 2001) as a minimum set that technical and author teams considered as context in preparing portions of this assessment. These two scenarios are referred to as the Special Report on Emissions Scenarios A2 and B1 scenarios, which assume either a continuation of recent trends in fossil fuel use (A2) or a vigorous global effort to reduce fossil fuel use (B1). Temperature increases from 4 to 22°F are predicted (to 2070–2099) depending on which emissions scenario (A2 or B1) is used with the least warming in southeast Alaska and the greatest in the northwest. Concomitant with temperature changes, by the end of the 21st century the growing season is expected to lengthen by 15–25 days in some areas of Alaska, with much of that corresponding with earlier spring snow melt.

Future projections of precipitation (30–80 years) over Alaska show an increase across the State, with the largest changes in the northwest and smallest in the southeast. Because of increasing temperatures and growing season length, however, increased precipitation may not correspond with increased water availability, due to temperature related increased evapotranspiration.

The extent of snow cover in the Northern Hemisphere has decreased by about 10 percent since the late 1960s, with stronger trends noted since the late 1980s. Alaska has experienced similar trends, with a strong decrease in snow cover extent occurring in May. When averaged across the State, the disappearance of snow in the spring has occurred from 4 to 6 days earlier per decade, and snow return in fall has occurred approximately 2 days later per decade. This change appears to be driven by climate warming rather than a decrease in winter precipitation, with average winter temperatures also increasing by about 2.5°F.

The extent of sea ice has been declining, as has been widely published in both national and scientific media outlets, and is projected to continue to decline during this century. The observed decline in annual sea ice minimum extent (September) has occurred more rapidly than was predicted by climate models and has been accompanied by decreases in ice thickness and in the presence of multi-year ice. This decrease was first documented by satellite imagery in the late 1970s for the Bering and Chukchi Seas, and is projected to continue, with the potential for the disappearance of summer sea ice by mid- to late century.

A new phenomenon that was not reported in previous assessments is ocean acidification. Uptake of carbon dioxide (CO_2) by oceans has a significant effect on marine biogeochemistry by reducing seawater pH. Ocean acidification is of particular concern in Alaska, because cold sea water absorbs CO_2 more rapidly than warm water, and a decrease in sea ice extent has allowed increased sea surface exposure and more uptake of CO_2 into these northern waters. Ocean acidification will likely affect the ability of organisms to produce and maintain shell material, such as aragonite or calcite (calcium carbonate minerals structured from carbonate ions), required by many shelled organism, from mollusks to corals to microscopic organisms at the base of the food chain. Direct biological effects in Alaska further along the food chain have yet to be studied and may vary among organisms.

Some of the potentially most significant changes to Alaska that could result from a changing climate are the effects on the terrestrial cryosphere—particularly glaciers and permafrost. Alaskan glaciers are changing at a rapid rate, the primary driver appearing to be temperature. Statewide, glaciers lost 13 cubic miles of ice annually from the 1950s to the 1990s, and that rate doubled in the 2000s. However, like temperature and precipitation, glacier ice loss is not spatially uniform; most glaciers are losing mass, yet some are growing (for example Hubbard Glacier in southeast Alaska). Alaska glaciers with the most rapid loss are those terminating in sea water or lakes. With this increasing rate of melt, the contribution of surplus fresh water entering into the oceans from Alaska's glaciers, as well as those in neighboring British Columbia, Canada, is approximately 20 percent of that contributed by the Greenland Ice Sheet.

Permafrost degradation (that is, the thawing of ice-rich soils) is currently (2012) impacting infrastructure and surface-water availability in areas of both discontinuous and continuous ground ice. Over most of the State, the permafrost is warming, with increasing temperatures broadly consistent with increasing air temperatures. On the Arctic coastal plain of Alaska, permafrost temperatures showed some cooling in the 1950s and 1960s but have been followed by a roughly 5°F increase since the 1980s. Many areas in the continuous permafrost zone have seen increases in temperature in the seasonally active layer and a decrease in re-freezing rates. Changes in the discontinuous permafrost zone are initially much more observable due to the resulting thermokarst terrain (land surface formed as ice rich permafrost thaws), most notable in boreal forested areas.

Climate warming in Alaska has potentially broad implications for human health and food security, especially in rural areas, as well as increased risk for injury with changing winter ice conditions. Additionally, such warming poses the potential for increasing damage to existing water and sanitation facilities and challenges for development of new facilities, especially in areas underlain by permafrost. Non-infectious and infectious diseases also are becoming an increasing concern. For example, from 1999 to 2006 there was a statistically significant increase in medical claims for insect-bite reactions in five of six regions of Alaska, with the largest percentage increase occurring in the most northern areas. The availability and quality of subsistence foods, normally considered to be very healthy, may change due to changing access, changing habitats, and spoilage of meat in food storage cellars.

These and other trends and potential outcomes resulting from a changing climate are further described in this report. In addition, we describe new science leadership activities that have been initiated to address and provide guidance toward conducting research aimed at making available information for policy makers and land management agencies to better understand, address, and plan for changes to the local and regional environment.

This report cites data in both metric and standard units due to the contributions by numerous authors and the direct reference of their data.

Introduction

Alaska's climate appears to be in a state of flux. Some patterns of change and associated consequences may be clear, such as the losses in sea ice, glaciers, and permafrost, whereas others are more subtle, such as the foothold that some invasive species have found in various parts of the State. Additional non-climate related changes are taking place in Alaska in a number of different sectors—the natural environment, energy production and use, and human social systems. Although human communities and natural systems are experiencing change from a variety of stressors, this report focuses specifically on climate related changes, effects, and societal consequences.

This Alaska Technical Regional Report (part of the National Climate Assessment) looks at current changes; synthesizes relevant and new science and information since publication of the last Alaska regional report (1999); and provides outlooks and projections of climate-related conditions (temperature, precipitation, snow cover, growing season, and permafrost extent). Like the 2013 National Climate Assessment that will be, in part, derived from the regional reports, our purpose is to increase the basic understanding of what is known and not known, in terms of the current and potential effects that a changing climate has and may have on water resources, transportation, ecosystems, human health, forestry, agriculture, and other socio-economic conditions. The information that this report provides is intended to help policymakers, land managers, and the general public become more informed about the current and potential effects of climate change in Alaska. The information also may assist in decision making that may reduce the overall vulnerability of the State and its people to a changing landscape.

The contents of this report are based on published scientific research that is in the public domain. This literature, most of which has been produced since publication of the 1999 Alaska Technical Regional Report, has been reviewed and summarized in this report by more than 40 contributing authors. Thus the information presented and the models shown have been either peer reviewed, or in the case of models, generally are accepted procedures that show potential future outcomes. This report also identifies important information needs and priority topics for continued monitoring and subsequent assessment activities that may be useful in conducting future regional climate assessments related to potential changing climate conditions. This report involved the contributions of many authors and the inclusion of pertinent scientific research, resulting in data presented throughout the report in both metric and standard units. The choice was made to maintain the author's original scientific units rather than to convert to one standard type of unit throughout the report, thus keeping the integrity of the information described here with that of the research citations used.

This report is organized into seven major sections following the introduction:

1. *Regional Description*: narratives that provide the geographic context, followed by a series of socio-economic summaries;

2. *Alaska's Climate Trends*: information about recent climate and climate-related phenomena whose influence is cross-cutting and which drive the changes described in other parts of the assessment;

3. *Regional Climate Forecast*: current and projected climate forecast and related environmental conditions;

4. *Observed Environmental Trends*: information about changes and influences of climate change on the ocean, hydrologic linkages, the land, and the human environment;

5. *Potential Effects of a Changing Climate*: summaries of the potential effects of a changing climate on Alaska;

6. *New Science Leadership on the Alaskan Landscape*: science leadership activities that have been initiated since the 1999 report was produced; and

7. *Planning for the Future*: suggested activities that could improve future assessments and better engage the public for input.

This 2012 National Climate Assessment – Alaska Technical Regional Report was produced through the guidance and direction of the report writing team (2012 NCA Alaska Technical Regional Report Writing Team). In addition to providing input to, and provide edits of, the text, each member also provided recommendations of contributing authors that could provide science summaries to each of the sections of the report.

2012 NCA Alaska Technical Regional Report Writing Team

- Carl Markon—U.S. Geological Survey (USGS; Team Lead)
- F. Stuart Chapin III—University of Alaska Fairbanks (UAF)– National Climate Assessment- National Climate Assessment and Development Advisory Committee, Federal Advisory Committee Act member
- Sarah F. Trainor—UAF/Alaska Center for Climate Assessment and Policy (ACCAP)
- Vanessa Skean—USGS
- Stephen Gray—Alaska Climate Science Center/USGS
- Michael Brubaker—Alaska Native Tribal Health Consortium
- Durelle Smith—USGS
- Philip Loring—UAF/ACCAP
- Jon Zufelt—U.S. Army Corps of Engineers
- Molly McCammon—Alaska Ocean Observing System
- James Partain—National Oceanic and Atmospheric Administration

Full contact information in provided in appendix A.

Regional Description

Geographic Context

Alaska, the largest State in the Nation, spans a land area of around 580,000 mi², almost one-fifth the size of the conterminous United States. Alaska is bounded by Canada (a border of 1,538 mi), the Arctic Ocean (Beaufort and Chukchi Seas), the Bering Sea, and the Pacific Ocean (fig. 2). Measured in a straight distance, Alaska's coastline is 6,640 mi in length; however, if each bay, fjord, island, and channel were considered, the length of the coastline of Alaska would be 33,555 mi, almost three times that of the conterminous United States. Point Barrow (71°23'N, 156°29'W) and Amatignak Island (51°15'N, 179°06'W) represent the State's northern and southern points, respectively. Alaska's eastern extreme is near Portland Canal in southeast Alaska (55°00'N, 130°00'W) and the western extreme is Cape Wrangell, Attu Island (52°55'N, 172°27'E).

Land ownership in Alaska can be divided into four groups: Federal, State, Native (private), and other private lands. The Federal Government is the largest landowner at 343,000 mi² (or 60 percent), which includes military reservations, national parks, national wildlife refuges, national forests, and the National Petroleum Reserve-Alaska. The State owns 164,000 mi² (or 28 percent) of the land, with the remainder in private Native corporation holdings (69,000 mi² or 12 percent) and those of other private individuals (<1 percent).

Alaska's Ecoregions

Alaska is not a uniform landscape. Combinations of climatic, geologic, edaphic (soil-related), hydrologic, vegetative, and other factors form ecological regions (ecoregions) of differing potentials to provide ecosystem services and capacities to buffer effects of change. The delineation of these regions provides an environmental framework useful for stratifying ecological variance for analysis, interpretation, resource management and reporting, and informing policy. The 1999 Alaska Technical Regional Report (Alaska Regional Assessment Group, 1999) cited a 1973 Ecosystems of Alaska (Joint-Federal State Land Use Planning Commission, 1973) to provide a general description of Alaskan ecosystem types; these ecosystems were, however, primarily vegetation-based and did not adequately represent the communities present and their interactions with the physical environment. Since that time, three new ecoregion frameworks have been developed for Alaska, representing different philosophies and approaches to delineating regions. Each framework is described here to show how new knowledge has increased the means by which ecoregions are described, ending in the most recent ecoregion map.

The earliest ecoregion map published for Alaska was part of a national scheme of ecoregions and subregions of the United States (Bailey and others, 1994). The regional delineation for Alaska was contributed by Nowacki and Brock (1995), following procedures that reflected the management perspective of the U.S. Forest Service, which viewed the framework as a basis for resource assessments, environmental analyses (for example, management feasibility and effects studies), watershed analyses, and scenario planning for future resource conditions and emerging concerns (Cleland and others, 1997). The framework included four levels of regional hierarchy, in which each successive level was delineated on the basis of a different environmental component (see Bailey, 1983, 1985, 1988): climatic characteristics were used to define the broadest level; potential natural vegetative features were highlighted at the next level; the distinction of montane versus non-montane terrain demarked the third level; and finer scaled physiographic characteristics were recognized at the fourth level (Bailey and others, 1994).

The second ecoregion map published for Alaska (Gallant and others, 1995) was based on the philosophy that the relative influence of formative regional drivers varies across space and cannot be prescribed for a consistent recipe to delineate regions. This approach contrasts with that followed by the U.S. Forest Service and was developed by the U.S. Environmental Protection Agency to derive general-purpose regions for environmental management (Omernik, 1987, 1995). The ecoregion framework has been used to understand patterns of land-cover and land-use change, evaluate physical, chemical, and biological characteristics of freshwater systems (for example, to develop biocriteria), and for wildlife conservation applications, among others (Omernik and others, 2011). Source data on climate, surficial and bedrock geology, physiography, hydrology, soils, permafrost, vegetation, and disturbance regimes in Alaska were consulted and variously emphasized to interpret key characteristics for delineating ecoregional boundaries. The resulting map provided a one-level, non-hierarchical structure and highlighted areas that were transitional between ecoregions. The ecoregion boundaries later were modified slightly along the international border to connect with Canadian ecoregions delineated in a similar fashion by Wiken (1986) as part of a North American ecoregion framework (Commission for Environmental Cooperation, 1997) that also included two coarser levels of regional hierarchy formed from aggregating the original ecoregions.

The third effort to delineate ecoregions for Alaska (fig. 3; Nowacki and others, 2001) sought to produce a framework agreed upon across agencies to improve their ability to communicate and manage information for a common set of units. The developers incorporated additional, more recent sources of information, field expertise from researchers from multiple organizations, and elements from the approaches used for the earlier two ecoregion maps as a means to bridge their differences. The resulting map provided two levels of hierarchy and has been used for reporting by the U.S. Forest Service (U.S. Department of Agriculture, Forest Service, 2001; Schulz, 2003), for interagency climate-scenario exercises (Murphy and others, 2010), for studying patterns of fire regimes (Hu and others, 2010a), and for documenting vegetation responses to climate change (Verbyla, 2008).

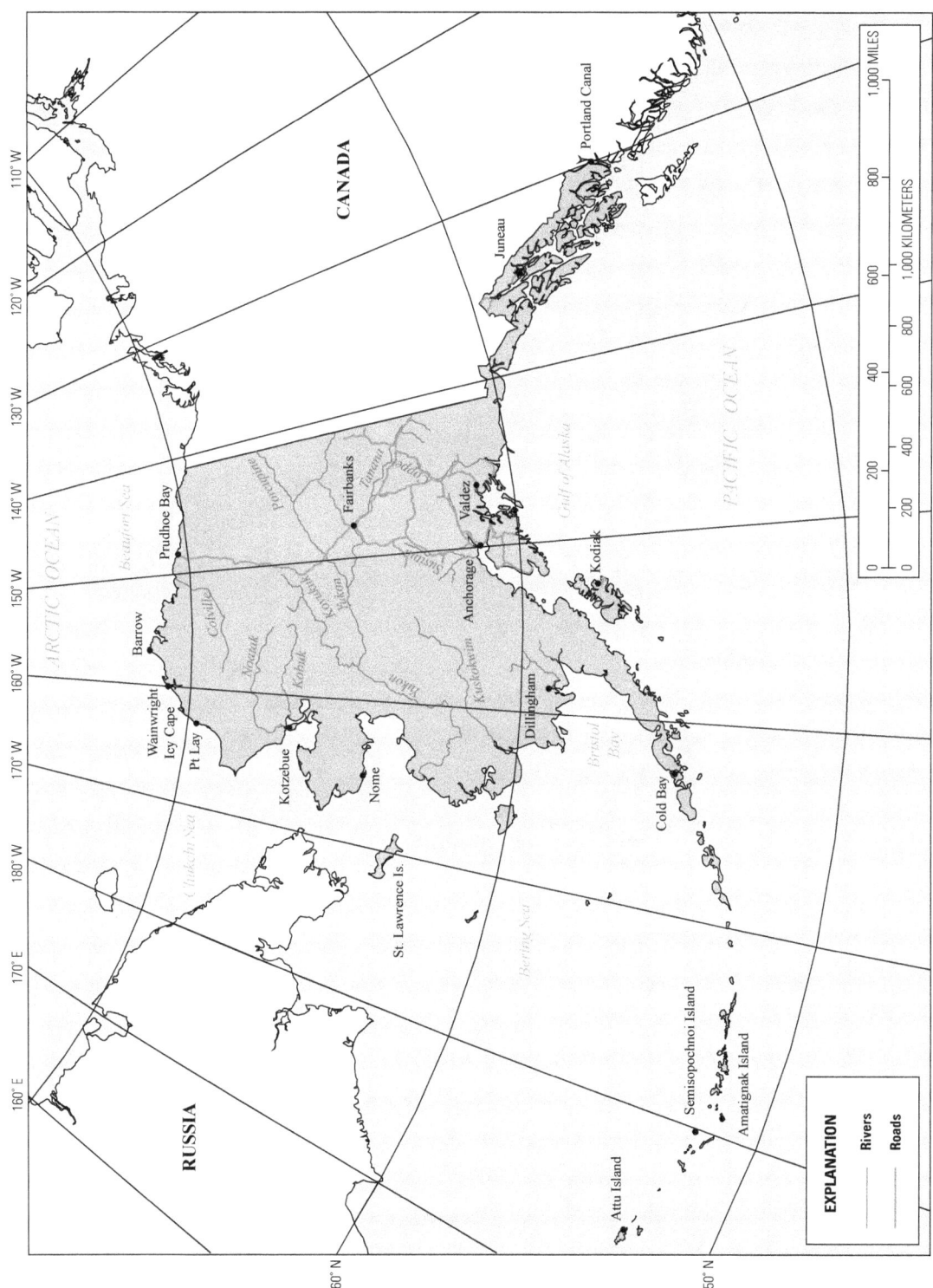

Figure 2. General geography of Alaska.

Ecoregions of Alaska and Neighboring Territories

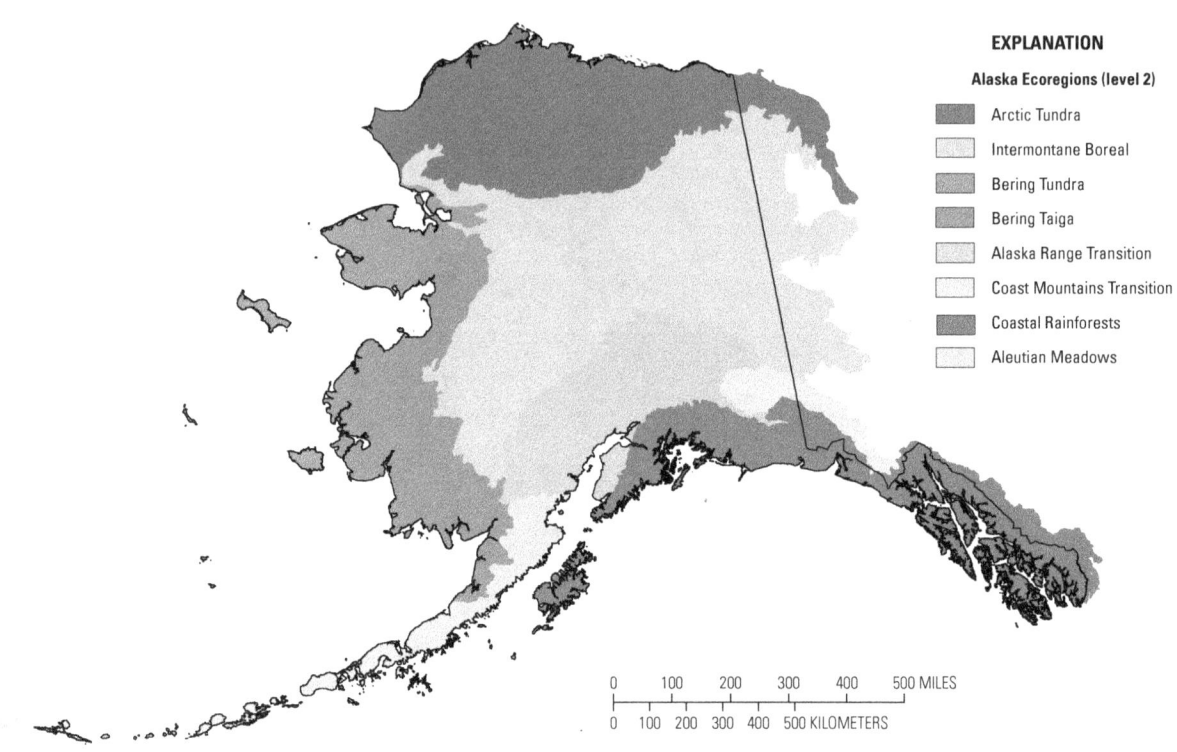

EXPLANATION

Alaska Ecoregions (level 2)

Arctic Tundra

Intermontane Boreal

Bering Tundra

Bering Taiga

Alaska Range Transition

Coast Mountains Transition

Coastal Rainforests

Aleutian Meadows

0   100   200   300   400   500 MILES

0   100   200   300   400   500 KILOMETERS

Figure 3. Unified ecoregions of Alaska (Nowacki and others, 2001).

On the basis of work by Nowacki and others (2001), the Alaskan ecoregions consist of three groups at the broadest scale (level 1) and eight groups at the moderate scale (level 2), establishing a baseline for studying, managing, and understanding the ecosystems of Alaska and their driving processes. The Level 1 ecoregions consist of Polar, Boreal, and Maritime. The key features of the Polar level are non-forested terrain with a cold, dry climate. It covers 24.8 percent of the land area of Alaska. The Polar level includes the Arctic Tundra and Bering Tundra groups in the Level 2 classification. The Boreal level consists of the Intermontane Boreal, the Bering Taiga, and Alaska Range Transition groups and is the dominant Level 1 group in that it covers 56.9 percent of the surface. This level is identified by a forested landscape with a dry climate, a large seasonal variation in temperature, and areas prone to forest fires. The third Level 1 group is the Maritime level. It is restricted to the southern portion of the State and it is the smallest of the Level 1 regions. It covers 18.3 percent of the land area and includes the Level 2 groups of Coastal Rainforests, the Coastal Mountains Transition, and the Aleutian Meadows. The Maritime groups are characterized by forested landscapes (with the exception of the Aleutian Meadows, which are treeless) with a warm, wet climate, limited variation in temperature throughout the year, and areas subject to high wind events.

Permafrost

Permafrost is a unique driver of ecosystem processes in Alaska. In addition to physically supporting ecosystems, permafrost controls soil temperature and moisture, subsurface hydrology, rooting zones and microtopography (Woo, 1992) and plays an important role in supporting infrastructure (U.S. Arctic Research Commission Permafrost Task Force, 2003). About 470,000 mi² (81 percent) of Alaska's land surface belong to the permafrost zone. Permafrost, defined as any ground with temperatures below 32°F for at least two consecutive years, is differentiated into four major subzones (based on Jorgenson and others, 2008):

1. In the continuous permafrost zone, 90–100 percent of the region is underlain by permafrost (32 percent of Alaska's land surface in the northern part of the State);

2. In the discontinuous zone, 50–90 percent of the region is underlain by permafrost (31 percent of Alaska's land surface in the south-central and interior part of the State);

3. In the sporadic zone, 10–50 percent of the region is underlain by permafrost (8 percent of Alaska's land surface in the southern part of the State); and

4. In the isolated zone, 0–10 percent of the region is underlain by permafrost (10 percent of Alaska's land surface also in the southern part of the State).

Permafrost temperatures in Alaska follow a south-to-north gradient as well as a low-to-high elevation gradient, with lower mean annual ground temperatures found on the Alaska North Slope (15°F) and at higher elevations in mountain ranges. Total permafrost thickness is highly variable and depends on factors such as climatic and geological history, and on regional-to-local factors such as lithology, geomorphology, vegetation, soils, and climate. Maximum permafrost thicknesses of 2,166 ft below surface have been recorded on the Alaska Arctic coastal plain near Prudhoe Bay. Above the perennially frozen layer of permafrost, a thin surface soil layer that seasonally thaws and freezes every year is termed the 'active layer'. The active layer ranges in thickness from 0.8 to 13 ft, with the thickest layers in more southern and warmer climate zones and in the mountains.

Permafrost may consist of any ground material, including bedrock, sediments, soils, organic material, and ground ice. Accordingly, the characteristics of permafrost are not uniform across Alaska and responses to environmental change can differ vastly (Jorgenson and others, 2010). The vulnerability of permafrost to warming and thaw depends on a complex set of permafrost properties and environmental variables. Warming air temperatures do not necessarily result in a linear warming of permafrost, but is complicated by factors that insulate the frozen ground, such as vegetation cover, soil organic layers, and snow; and other factors that increase heat transfer into the ground, such as ground water, lakes and rivers. Accordingly, changes in these factors due to natural or anthropogenic disturbances have an effect on the ground thermal regime and thus the underlying permafrost.

Socio-Economic Conditions

The population of Alaska in 2010 was estimated to be 710,231 (U.S. Census Bureau, 2010). According to the State of Alaska, Division of Community Advocacy, the population is dispersed throughout 358 communities. Anchorage is the largest urban setting, with a population of 291,826 in 2010 (U.S. Census Bureau, 2010). The State population is expected to gradually increase throughout the 21st century as areas outside of Anchorage continue to be developed.

Three major commodities—oil and gas, minerals, and seafood—contribute most to the economy of Alaska. Tourism is a strong fourth part of the economy, and farming is a minor contributor.

Oil and gas makes up approximately 25 percent of the State's Gross Domestic Product, and in 2010 represented 4 percent of all wage and salary employment in Alaska (Fried, 2011). Most all of the oil industry work force is in three areas: Anchorage, which houses the industries headquarters; and the North Slope and Kenai Peninsula, where the oil is produced. Oil and gas in Alaska accounted for 44,800 jobs and just under $2.65 billion in annual payroll to Alaska residents in 2010, including all direct, indirect, and induced employment and

wages. This accounted for approximately 10 percent of all employment in Alaska and 13 percent of all resident earnings (McDowell Group, 2011a).

Minerals have recently become the second most valuable commodity in the State, with the most important economic minerals mined, in order of importance, being zinc, gold, lead, coal, sand and gravel, building stone, silver, copper, and jade. In 2011, gross mineral production value was $3.8 billion that contributed 4,500 direct industry jobs and $148 million paid to State Government revenue through rent, royalties, fees, and taxes (Alaska Miners Association, Inc., 2008). Mineral exploration expenditures in Alaska during 2010 were at least $264.4 million, with approximately 70 percent coming from Canadian sources. Exploration occurred all across the State, but more than $127 million (or 48 percent of the exploration funds) was spent in southwestern Alaska and $55 million was spent in the eastern interior region. Thirty-four projects reported exploration expenditures of $1 million or more and 47 additional projects expended at least $100,000 (Szumigala and others, 2011).

Commercial fisheries constitute the third largest industry in Alaska, behind oil and gas and mining (Northern Economics, 2009). Alaska's commercial fisheries account for roughly 50 percent of the United States' total wild fish landings and led all States in terms of both volume and ex-vessel value of commercial fisheries landings in 2009 (1.84 million metric tons worth $1.3 billion dollars, National Marine Fisheries Service, 2010; fig. 4). In the

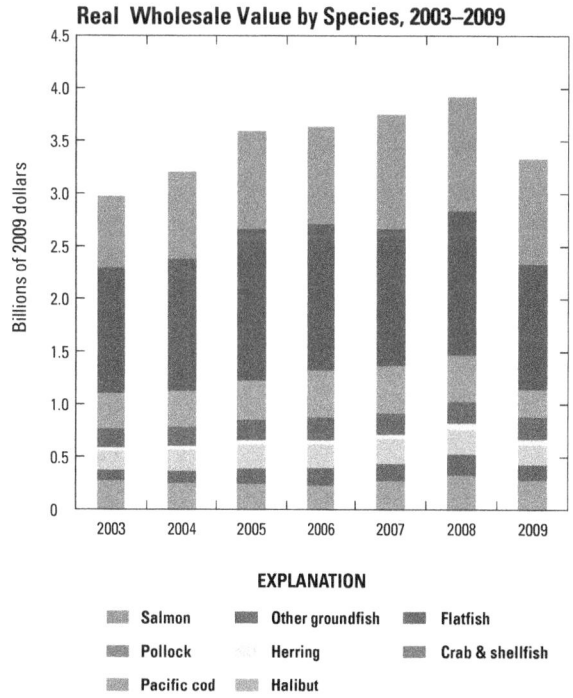

Real Wholesale Value by Species, 2003–2009

EXPLANATION

- Salmon
- Other groundfish
- Flatfish
- Pollock
- Herring
- Crab & shellfish
- Pacific cod
- Halibut

Figure 4. Real wholesale value for Alaska commercial fisheries stocks (Northern Economics, 2011).

list of top 50 U.S. ports based on volume for 2009, Alaska had 11, including Dutch Harbor-Unalaska (1); Kodiak (4); Naknek-King Salmon (11); Sitka (14); Ketchikan (15); Petersburg (18); Cordova (21); Seward (26); Homer (36); Juneau (41); and Kenai (50). These fisheries create $5.8 billion in direct and indirect output, and employ more workers than any other industry sector in Alaska, with 78,519 direct and indirect workers (Northern Economics, 2009). In addition, thousands of Alaskans and visitors fish recreationally in Alaska waters. Rural Alaska Native communities increasingly benefit from commercial fisheries; 65 Bering Sea communities, for instance, participate in a Community Development Quota (CDQ) program designed to bolster rural participation in these rationalized commercial fisheries. From 1992 through 2008, the CDQ Program generated more than $240 million in wages, payments to participants, and scholarships and training benefits (Western Alaska Community Development Association, 2008); likewise in 2008, CDQ entities provided wage and salary jobs to more than 1,600 individuals, and the combined payroll for the year exceeded $22.3 million (Western Alaska Community Development Association, 2008).

Tourism has been one of the fastest growing contributors to the State's economy since the 1990s, with annual visits increasing from roughly 1.1 million in 1994 to more than 1.7 million in 2004 (Northern Economics, 2004). The increase in numbers of visitors statewide is reflected in the number of jobs created in the tourism industry, which added more jobs than any other basic industry through the 1990s (Leask and others, 2001). The number of visitors to Alaska peaked in the 12-month period between May 2007 and April 2008 (nearly 1.95 million) but began declining after that. The most recent estimate of visitors is 1.75 million in 2010–11 (McDowell Group, 2011b; 2011c).

The leading farming regions of Alaska are the Matanuska Valley, northeast of Anchorage, and Delta Junction, south of Fairbanks. The short growing season and expense of getting agricultural products to market are limiting factors, but seasonal open-air markets are common and often well attended. In 2002, Alaska had 590 farms covering 920,013 acres, with hay, potatoes, lettuce, cabbage, carrots, beef, pork, dairy products, and greenhouse and nursery items being common commodities, with a total value exceeding $4 million (City-Data, 2012).

Forests cover one-third of Alaska (Parson and others, 2001) and are regionally and globally significant (Wolken and others, 2011). Ninety percent of Alaskan forests are classified as boreal, representing 4 percent of the world's boreal forests (Shvidenko and Apps, 2006), and are located throughout interior and south-central Alaska. The remaining 10 percent of Alaskan forests are classified as coastal-temperate, representing 19 percent of the world's coastal-temperate forests (National Assessment Synthesis Team, 2003), and are located in southeast Alaska (fig. 5).

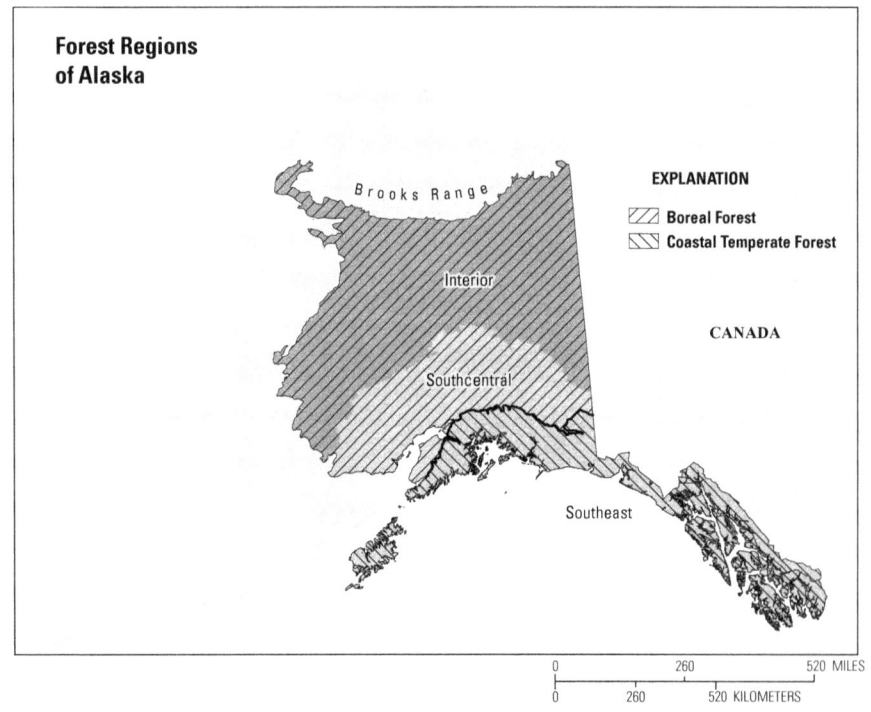

Figure 5.   Map of the boreal (interior and south-central) and coastal-temperate (southeast) forest regions of Alaska.

Capacity of Alaska's Fishing Communities to Cope with Change

Climate change is anticipated to have a variety of direct and indirect impacts on fish populations of commercial importance to Alaska. Berns (2010) developed a framework for the rapid assessment and visualization of risk to five Alaska's fishing communities: Cordova, Kodiak, Petersburg, Seward, and Sitka, all situated around the Gulf of Alaska. The assessment evaluates the relative risk of climate change impacts on fishing-dependent communities, first by using a productivity-susceptibility analysis method developed by the National Oceanic and Atmospheric Administration (Patrick and others, 2010) for each major commercially fished species, and then weighing these results to the relative contribution of each species to total catch for each community. Next, these determinations of risk are incorporated into an assessment of community capacity to respond to impacts using indicators built from existing socioeconomic data. The results are then visualized with 3-axes polygons to provide an overall impression of vulnerability, and to communicate the relative capacity of each community to respond to climatic change.

Social, economic, and natural capitals (assets) are ranked on a scale of 1 (most vulnerable) to 3 (most adaptable), when coping with potential climate change. In the figure below, the larger the region displayed, the less vulnerable the community is to change. Kodiak shows the greatest capacity for responding to climate change. Petersburg shows a low overall capacity, with social capital being less affected than economic and natural capital; Sitka has moderate economic capacity with which to respond to climate-driven impacts on fisheries but lacks a comparable level of social and natural capital, while Cordova and Seward are strongest in respect to natural capital.

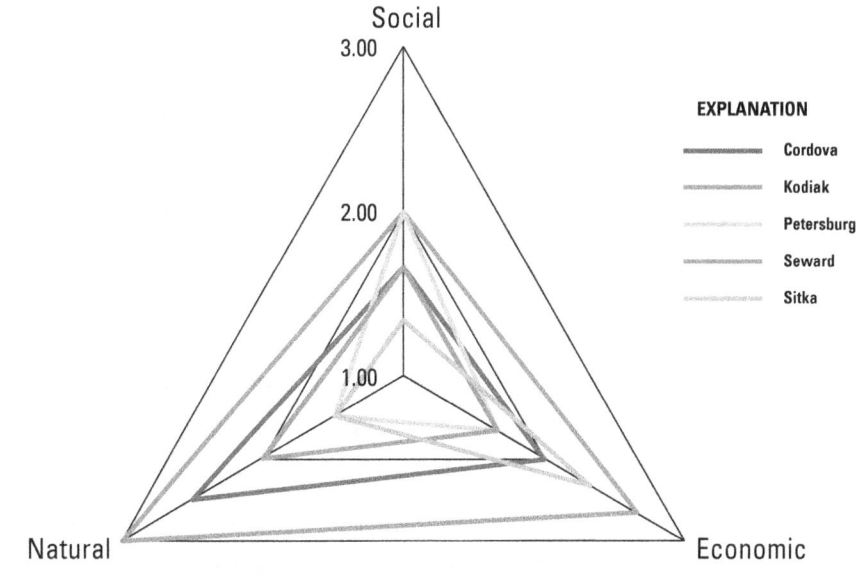

Tourism

Cruise ships play a major role in the State's tourism industry (Cerveny, 2004; Colt and others, 2007; Ringer, 2010) and contribute to important seasonal and regional variability in visitor numbers and spending (McDowell Group, 2010, 2011a, 2011b). For example, in the summer of 2009, 65 percent of the nearly 1.6 million visitors to Alaska came by cruise ships. Of the $1.3 billion in 2009 visitor spending, 39 percent occurred in the southeast region and 39 percent occurred in south-central region; the other 22 percent was split between the southwest, interior, and far northern regions. In addition to the normal tourist attractions, researchers think tourists are increasingly seeking out attractions that are vulnerable to climate change, such as this glacier in southeast Alaska.

Photograph by Mr. Bill Vanderford.

There are two National Forests in Alaska: the Tongass and the Chugach. The Tongass National Forest encompasses 26,000 mi², of which 15,400 mi² are forested and 5,600 mi² are deemed harvestable and available to commercial activities, with the target species being Sitka spruce, western hemlock, western red cedar, and yellow cedar. The Chugach National Forest spans 9,300 mi². Currently (2012), none of the land in the Chugach National Forest is classified as suitable for timber production (U.S. Department of Agriculture, Forest Service, 2002). Commercial logging began in southeast Alaska during the 1920s and boomed in the Tongass National Forest during the 1950s where the industry dominated. Historically, the timber industry grew through the 1980s, but by the late 1990s, the reduced harvests led to closure of pulp mills, which caused a 50-percent reduction of the work force (Leask and others, 2001; Beier and others, 2009).

Logging also has occurred on the Kenai Peninsula in response to the spruce bark beetle outbreak that caused extensive tree mortality in the mid-1990s. The spruce bark beetle infestation on the Kenai Peninsula was considered to be the most extensive insect infestation in North America at that time. Since 1989, the U.S. Forest Service and the Alaska Division of Forestry have mapped more than 138,000 acres of beetle-infected forests on the Kenai Peninsula (and more than 6 million acres statewide; Alaska Department of Natural Resources, Division of Forestry, 2010). The infestation of the 1990s accelerated in 1992 and peaked in 1996; however, new stands continue to be infested each year. The Kenai Peninsula Borough, Spruce Bark Beetle Mitigation Program (Kenai Peninsula Borough, 2012) identified 133,000 acres of harvested forest on the Kenai by using satellite imagery from 2000 to 2005. The boreal forests of the interior have been logged to a much lesser degree, but there is increased interest in harvesting resources there. The target species in the boreal forest are white spruce, quaking aspen, and paper birch.

Alaska's Climate Trends

Alaska's climate is influenced by three main factors: latitude, altitude, and geographic location, including seasonal distribution of sea ice (Alaska Climate Research Center, 2009). The State's vast expanse and geographical variation lead to a variety of climate types:

1. The southeast, south coast and southwestern islands experience a maritime climate with high precipitation and moderate temperatures. The State's highest annual average temperatures and highest precipitation amounts are found in this region (Shulski and Wendler, 2007; Western Regional Climate Center, Desert Research Institute, 2011).

2. In west-central Alaska, seasonal distribution of sea ice plays a major role in the regional climate. It experiences a maritime influence until sea ice forms in the Bering Sea, leaving the west coast with a more continental-like climate. Sea ice generally is established along the coast by late fall and remains until late spring (Shulski and Wendler, 2007; Western Regional Climate Center, Desert Research Institute, 2011).

3. A transitional zone between maritime and continental climates exists in the western part of Bristol Bay, southern Cook Inlet, and the southern part of the Copper River basin. This transitional zone is largely cutoff from maritime influence by mountains, but experiences moderate temperatures in comparison to the continental interior climate (Shulski and Wendler, 2007; Western Regional Climate Center, Desert Research Institute, 2011).

4. Interior Alaska, bounded by the Brooks and Alaska Ranges, experiences a truly continental climate, with large annual temperature variability, low humidity, and relatively light and irregular precipitation. Summers are warm and sunny, while winters are long and cold, with frequent low-level temperature inversions caused by radiational cooling at the surface (Shulski and Wendler, 2007; Western Regional Climate Center, Desert Research Institute, 2011).

5. North of the Brooks Range is the Arctic region of Alaska, where the State's lowest annual average temperatures are found. Bordered on the north by the Arctic Sea, coastal areas of this region are affected by the moderating effect of sea ice free seasons. Summers are cool and cloudy along the coast, and temperatures are more continental farther inland with warmer summers and cooler winters. Precipitation is relatively light, but commonly under-reported, as frequent high winds result in gauge undercatch; blizzard conditions may be common in the winter (Shulski and Wendler, 2007).

Historical and current climate conditions are summarized for temperature and precipitation in Alaska, including coastal storm events.

Temperature

The most recent, comprehensive and statistically rigorous analyses of Alaska's climate records indicate that average-annual statewide temperatures have increased by nearly 4°F over the period 1949–2005 (Stafford and others, 2000; Shulski and Wendler, 2007). This level of warming, however, is not consistent across seasons. The bulk of observed temperature increases occurred over winter (+6.3°F) and spring (+4.1°F). Statewide temperatures during the months of summer and fall have risen by only 2.3 and 1.4°F, respectively.

Historical temperature change within Alaska also varies by region and observing site. On the basis of data from 1949 to 1998, increases in mean annual temperature have been greatest (3.9°F) in the State's interior (Stafford and others, 2000). Generally speaking, southwestern Alaska has seen the smallest increases in average annual temperature, with changes on the order of 1.8–2.5°F. Examination of individual station records further highlights differences in warming trends between interior and more coastal or maritime locations. Mean annual temperatures for Fairbanks in the State's interior have risen by 8.2°F over the period 1949–2005, compared to a 1.6°F increase over the same period for St. Paul Island in the Bering Sea (Shulski and Wendler, 2007; Wendler and Shulski, 2009).

A significant part of the observed warming in Alaska occurred as a sudden, step-like change in the mid-1970s. With the exception of Barrow, this step change has been documented at all of the State's first-order stations, and is reflected in statewide averages (Shulski and Wendler, 2007;

Wendler and Shulski, 2009). The mid-1970s step change coincides with a major shift in atmospheric circulation patterns across a large portion of the Pacific basin, called the Pacific Decadal Oscillation (PDO). The PDO index captures this shift as a transition from predominantly negative to predominantly positive values around 1976–1977 (Mantua and others, 1997). Historically speaking, a positive PDO is associated with a strong Aleutian low, which serves to direct warm air into interior Alaska (Wendler and Shulski, 2009). Warm-air advection associated with a positive PDO is especially prevalent in the winter months. In turn, it is very likely that some portion of observed 20th century warming in Alaska can be attributed to inherent decadal-scale variability in regional climate. The temperature increase in Alaska, however, mirrors trends across the Arctic and sub-Arctic (Hinzman and others, 2005; Solomon and others, 2007) suggesting that PDO-like variability may have amplified or accelerated an underlying long-term warming trend.

Increasing mean annual or seasonal temperatures also are associated with significant changes in temperature extremes. As above, the magnitude of observed changes varies by season and location, but general trends in Alaska point toward increasing minimum and maximum temperatures, as well as fewer days of extreme cold (Shulski and Wendler, 2007). Extreme high temperatures also show a positive trend. Examination of the greater than 100-year climate record from Fairbanks, for example, shows how the frequency of days below –40°F has gone from roughly 14 to 8 days per year over the past century (Wendler and Shulski, 2009). Based on this same record, the number of days above the freezing point (32°F) has increased from 85 to 123 over the last 100 years.

Coincident with a known phase shift in the Pacific Decadal Oscillation from its predominant negative (cold) phase to predominant positive (warm) phase in 1976, there has been a clear decrease in the occurrence of 4-day and 7-day cold-wave events. Although not as readily apparent as the decrease in occurrence of cold waves, the average heat wave index after 1976 is three times that of the average value seen prior to 1976 (Alaska Climate Research Center, 2009).

Warm and cold temperature extremes in Alaska display similar regional and seasonal variation as that of mean temperatures. In a study examining the records for 26 Alaskan observing stations from 1950 to 2008, the greatest increase in frequency of warm extremes (warmest 1 percent of daily high temperatures) and the greatest decrease in frequency of cold extremes (coldest 1 percent of daily lows) is found in spring (table 1). The observed decrease in frequency of extreme cold events in winter has been more pronounced in the past few decades. The next greatest increases (decreases) in warm (cold) extremes are observed in winter. Figures 6 and 7 show values of heat wave and cold wave indices, respectively, for the State of Alaska based on data obtained from the same 26 stations as for table 1.

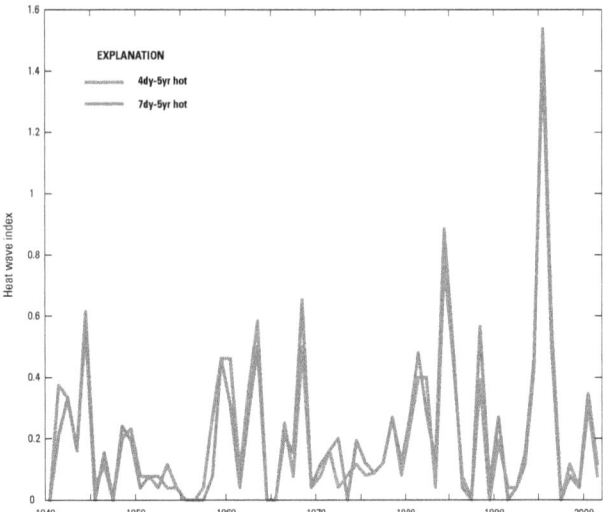

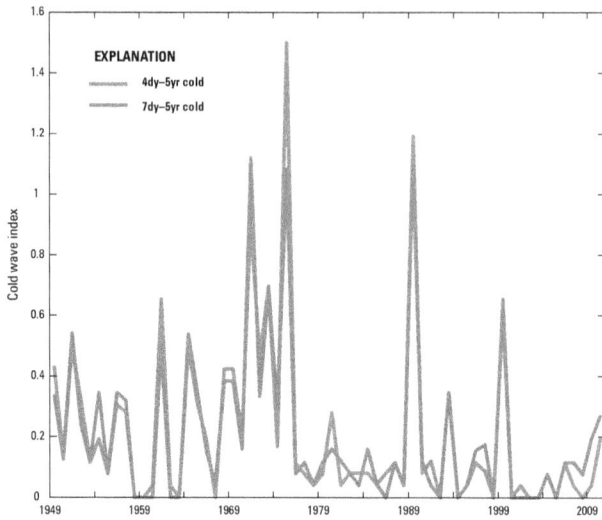

Figure 6. Heat wave indices for Alaska. Time series (1949–2010) of an index for the occurrence of heat waves defined as 4-day periods (blue) and 7-day periods (red) that are warmer than the threshold for a 1-in-5-year recurrence. Based on data from the National Climatic Data Center Cooperative Observer Network and based on methods from Kunkel and others (1999).

Figure 7. Cold wave indices for Alaska. Time series (1949–2010) of an index for the occurrence of cold waves defined as 4-day periods (blue) and 7-day periods (red) that are colder than the threshold for a 1-in-5 year recurrence. Based on data from the National Climatic Data Center Cooperative Observer Network and based on methods from Kunkel and others (1999).

Table 1. Percentage of stations in each region in Alaska displaying upward trends in occurrence of warm extremes and downward trends in frequency of cold extremes in spring (1950–2008).

Region	Warm extremes (percent)	Cold extremes
Arctic[1]	100	100
West central	100	100
Interior	100	100
Southwest	100	100
South central	100	100
Southeast	71	86

[1] Only one station in the Arctic region (Barrow) had sufficient data for this analysis.

Precipitation

Trends in precipitation show more variability than those for temperature, in part because of the difficulties in measuring precipitation (Curtis and others, 1998), but primarily because environmental factors have a greater influence on precipitation than on temperature, which tends to follow more regional or synoptic-scale patterns. Missing or incomplete records further complicate the analysis of precipitation trends (Wendler and Shulski, 2009).

On the whole, Alaska saw a 10-percent increase in statewide average precipitation over the period from 1949 through 2005 (Shulski and Wendler, 2007). The greatest increases in precipitation were recorded in winter, with portions of western, southern, and most southeast Alaska showing precipitation increases of more than 20 percent for December–February (Stafford and others, 2000). Over the last five to six decades, summer precipitation decreased or stayed near long-term averages in much of the State.

Observations from Arctic stations and locations in far southeast Alaska show important exceptions to this increasing annual winter precipitation trend. Located on the central Arctic coast, Barrow showed a 36-percent decrease in precipitation over the period 1949–1998 (Curtis and others, 1998; Stafford and others, 2000). Likewise, in southeast Alaska, Annette showed a 24-percent decrease in average annual precipitation over the period 1949–2005 (Shulski and Wendler, 2007). Although not statistically significant, precipitation also decreased by 11 percent at Fairbanks over the period 1916–2005 (Wendler and Shulski, 2009).

As with average precipitation, the occurrence of extreme precipitation events is highly variable, both regionally and seasonally. Here, extreme precipitation events are defined as the heaviest 1 percent of 3-day precipitation totals for each calendar season. Based on data for 26 stations across Alaska, the most significant increases in extreme precipitation events are observed in the southeast and west-central regions in spring, although the Arctic region shows statistically significant decreases in the occurrence of extreme precipitation in all seasons except fall (table 2). Results (increases or decreases in precipitation) for summer and fall were largely insignificant with the exception of those for the southeast in fall, which showed significant increases for all but one station. With the exception of the Arctic, all regions have seen an increase in occurrence of extreme precipitation events in summer in the last few decades, although the statistical significance of this result has not been tested (Stewart, 2011).

Table 2. Percentage of stations in each region in Alaska displaying upward trends in occurrence of extreme 3-day precipitation events from 1950 to 2008.

Region	Winter (percent)	Spring (percent)	Summer (percent)	Fall (percent)
Arctic[1]	0	0	0	0
West central	67	100	0	67
Interior	33	50	50	50
Southwest	50	25	25	75
South central	40	40	40	60
Southeast	71	57	43	57

[1]Only one station in the Arctic region (Barrow) had sufficient data for this analysis.

Coastal Storm Events

An important societal concern is that extreme wave events have eroded inhabited shorelines of the western Alaskan coast, particularly in the southeast Chukchi Sea region (Francis, 2011). Although the waves created by these storms are not particularly high (10–13 ft), their effect on the low-lying permafrost-rich bluffs can cause several feet of shoreline to be washed away by one large storm.

Aleutian lows formed in the Gulf of Alaska and Bering Sea, and also bordering low-pressure and high-pressure systems, cause most of the storm activity that creates the largest wind events and the most extreme wave heights, which, in turn, cause the most damage in the western Alaskan region. These ocean waves generally are classified as wind-sea (that is, waves under the influence of winds in a generating area) and originate from the easterly direction. Swells (that is, waves moving away from a generating area and no longer influenced by winds) also are common but are at a much lower wave height (3 ft) and originate from the westerly direction, from a more open fetch area. Although less damaging than wind-sea, swell has a cumulative effect whose impacts also are felt (Francis, 2011).

Satellite altimeter radar data for the southeastern Chukchi Sea show a linear increase in mean significant wave height over the last two decades, with an average rate of 0.066 ft/yr (fig. 8; Francis and others, 2011). In the Pacific-Arctic region, the significant wave height has increased at a faster rate (0.082 ft/yr). The increase in wave height may be due to a longer open-water season owing to sea ice decline.

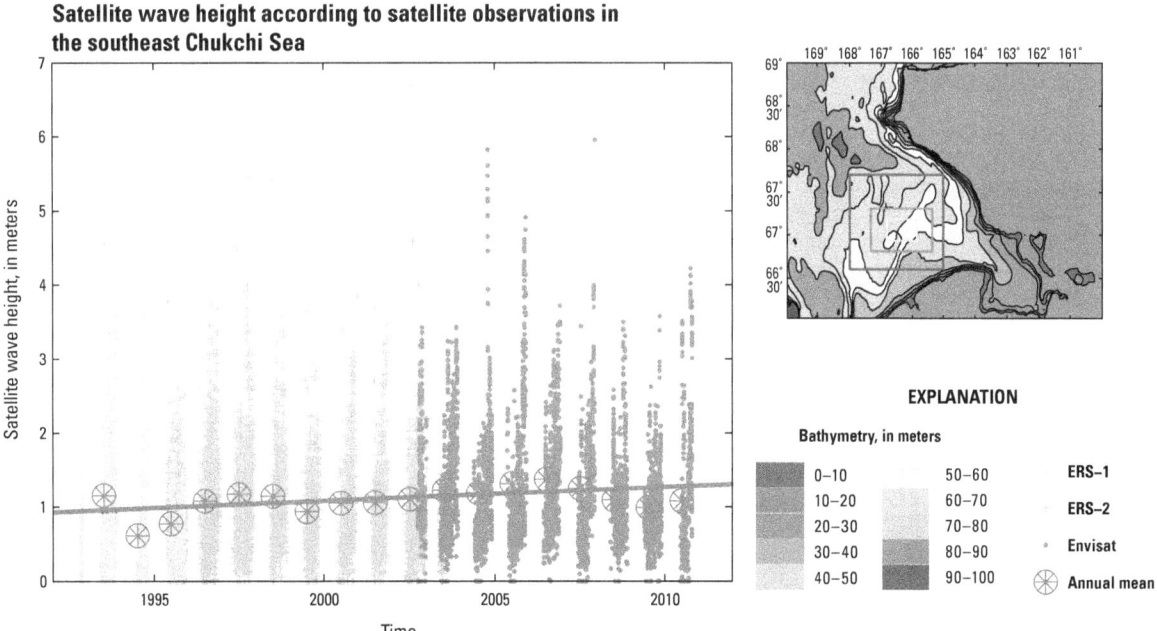

Figure 8. Satellite wave height observation from European Remote Sensing (ERS) Satellites-1/2 and Envisat in the southeastern Chukchi Sea from 1993 to 2010. Inset (top right) figure shows the domain (largest blue square) for all satellite wave height (SWH) observations for each year (yellow/green/red dots), and the mean average SWH for each year (blue wheels).

Regional Climate Forecasts

This initial forecast for the National Climate Assessment (NCA) Alaska region is based on the Coupled Model Intercomparison Project Phase 3[1] (CMIP3) General Climate Models dataset, in which 15 models identified in the 2009 NCA report were used (Karl and others, 2009a). The resolution of these datasets is 2.8 degrees latitude by 2.8 degrees longitude. The information provides statistics for the periods of 2021–2050, 2041–2070, and 2070–2099, with changes calculated with respect to the historical climate reference period of 1971–2000. Four different types of analyses are represented, as follows:

1. Multi-model mean maps.—Each model's data are first re-gridded to a common grid. Then, each grid point value is calculated as the mean of all available models' values at that grid point. Finally, the mean grid point values are mapped. Although this type of analysis weights all models equally, a number of research studies have found that the multi-model mean is superior to any single model in reproducing the present-day climate. A multi-mean analysis of future spatial patterns may be the most robust estimate of future change.

2. Spatially averaged products.—All grid point values within the Alaska region boundaries are averaged and represented as a single value. This is useful for general comparisons of different models, periods, and data sources. Because of the spatial aggregation, this product may not be suitable for many types of impacts analyses.

3. Probability density functions.—Spatially averaged values are calculated for each model simulation and are used to illustrate the differences among models. Frequency distribution of these spatially averaged values is displayed. This product provides an estimate of the uncertainty of future changes.

4. Downscaled temperatures, precipitation and growing season length.—These values were generated using a Delta[2] method applied to output from simulations run with two emissions scenarios (A2 and B1). As with the coarse model output, the downscaled results are presented as multi-model means for various timeslices.

Temperature Projections

The spatial distribution of the 15 CMIP3 multi-model mean annual temperature for Alaska is shown in figure 9 for three future time periods (2021–2050, 2041–2070, and 2070–2099) and two emissions scenarios A2 and B1. (Nakicenovic and others, 2000; Solomon and others, 2007). The simulation results for all three periods indicate an increase in temperature compared to that in 1971–2000. Southeast Alaska shows the least amount of warming, and the greatest temperature changes are in the far northwest.

Warming increases over time, as well as between scenario A2 and B1 for each respective period. Spatial variations are relatively small, especially for the B1 scenarios. For 2021–2050, B1 values range between 0 and 4°F and A2 values range slightly higher, from 0 to 6°F. For 2041–2070, warming in B1 is between 2 and 6°F and for A2 is from 2 to 8°F. Increases by 2070–2099 are larger still, with a 2–8°F range for B1 and a 4–9.5°F range for A2.

The mean annual temperature changes for each future time period and both emissions scenarios, averaged over the entire Alaska region for the 15 CMIP3 models, are shown in figure 10. The plus signs are values for each individual model, and the circles depict the overall means. Temperature changes increase over time for both scenarios, with greater changes for A2. By 2070–2099, increases for the B1 scenario are around 4.7°F, and for A2 are almost double, at 8.1°F.

A key overall feature is that early in the 21st century, the multi-model mean temperature changes are relatively insensitive to the emissions path, but late 21st century changes are quite sensitive to the emissions path, as indicated by the wider variance in the range of individual model changes than earlier in the century. There is considerable overlap between simulation results for the A2 and B1 scenarios, even for 2070–2099.

Figure 11 shows the mean seasonal changes for each future time period for the A2 scenario, averaged over the entire Alaska region for the 15 CMIP3 models. For all seasons, warming increases with time, with changes being smallest for spring. Temperature changes are comparable for the other seasons, ranging from around 3°F in 2021–2050 to more than 8°F in 2070–2099. The spread of individual model values is large in all cases, and also increases with time.

---

[1]Recent model outputs collected by the Program for Climate Model Diagnosis and Intercomparison (Lawrence Livermore National Laboratory). It is meant to serve Intergovernmental Panel on Climate Change's Working Group 1, which focuses on the physical climate system—atmosphere, land surface, ocean, and sea ice.

[2]An approach developed for bias removal of climate change assessments that assumes that future model biases for both mean and variability will be the same as those in present-day simulations.

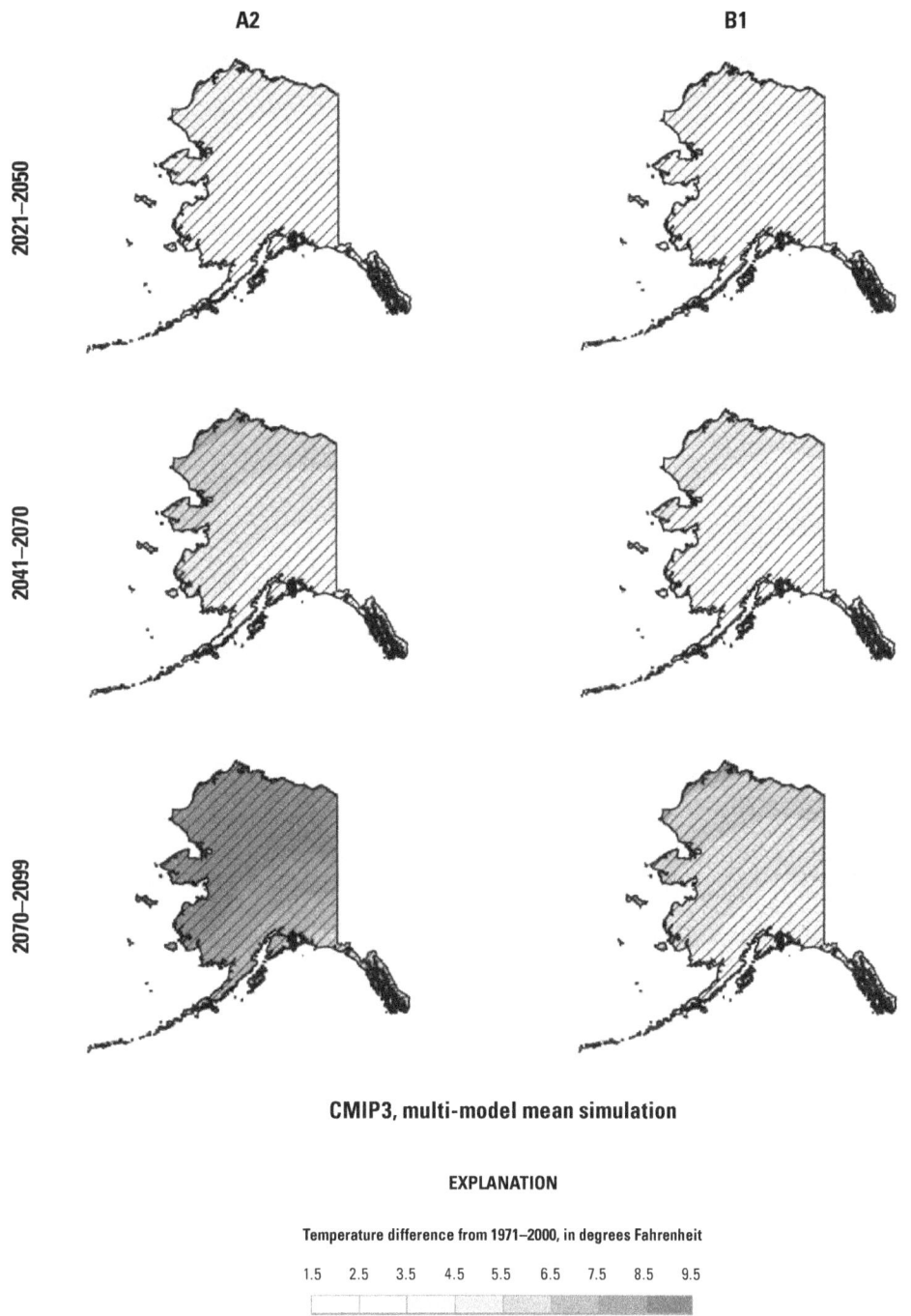

CMIP3, multi-model mean simulation

EXPLANATION

Temperature difference from 1971–2000, in degrees Fahrenheit

1.5   2.5   3.5   4.5   5.5   6.5   7.5   8.5   9.5

Figure 9. Multi-model mean annual differences in temperature (°F) between the three future periods and 1971–2000, from 15 CMIP3 model simulations. Areas with hatching indicate that more than 50 percent of the models show a statistically significant change in temperature. Map created by Michael Wehner, Lawrence Berkeley National Laboratory, 2011. CMIP3: Coupled Model Intercomparison Project Phase 3; A2: Intergovernmental Panel on Climate Change emissions scenario that assumes a continuation of recent trends in fossil fuel use; B1: Intergovernmental Panel on Climate Change emissions scenario that assumes a vigorous global effort to reduce fossil fuel use.

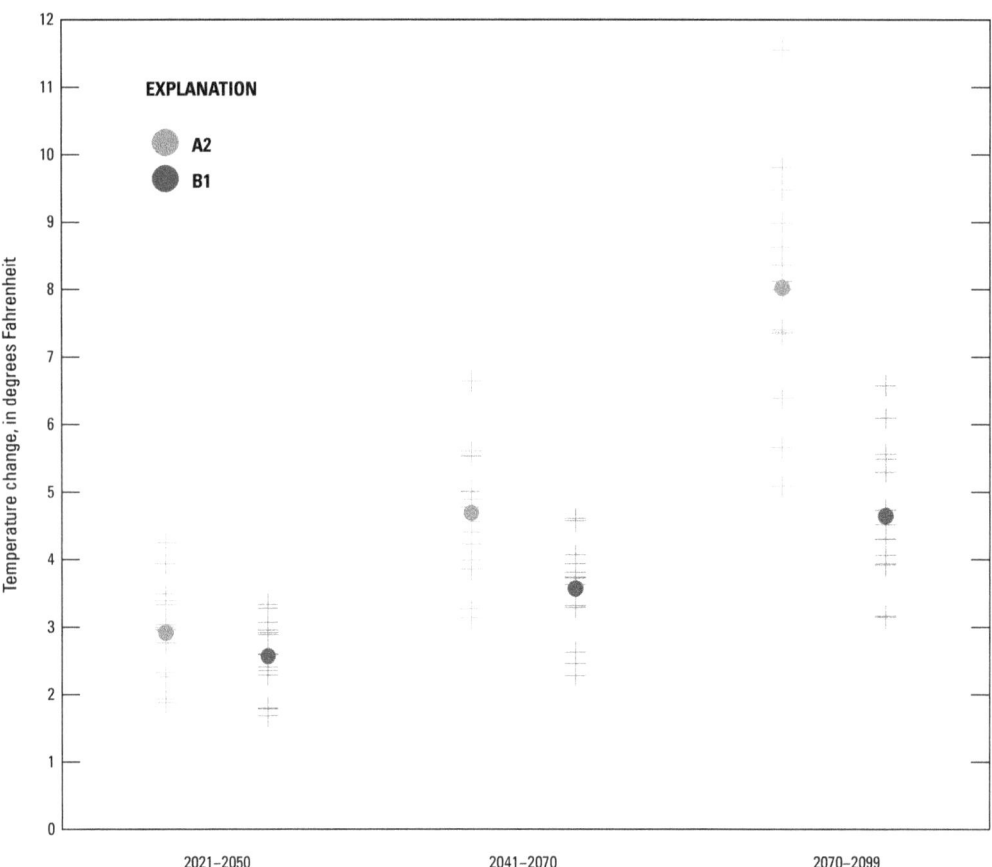

Figure 10.   Mean annual temperature changes (°F) for each future time period with respect to the reference period of 1971–2000 for all 15 CMIP3 models, averaged over the entire Alaska region for the high (A2) and low (B1) emissions scenarios. The plus signs are values for each individual model and the circles depict the overall means. CMIP3: Coupled Model Intercomparison Project Phase 3.

The distribution of changes in mean annual temperature for each future time period and both emissions scenarios across the 15 CMIP3 models is shown in table 3. The range of changes from lowest to highest ranges from 1.7°F in 2021–2050 for the B1 scenario to 11.6°F in 2070–2099 for the A2 scenario. The inter-quartile range of changes across the CMIP3 models (not shown) is between 0.6 and 1.4°F. Although the total range is seen to increase for each future time period, the inter-quartile range varies little.

Table 3 also illustrates the overall uncertainty arising from the combination of model differences and emission pathway. For 2021–2050, the projected changes range from 1.7 to 4.3°F and arise almost entirely from model differences. By 2070–2099, the range of projected changes has increased to 3.1 to 11.6°F, with roughly equal contributions to the range from model differences and emission pathway uncertainties.

Table 3.   Distribution of changes in mean annual temperature (°F) for the Alaska region for the 15 CMIP3 models.

[CMIP3, Coupled Model Intercomparison Project Phase 3; A2 and B1, from page 15]"

Scenario	Period	Low	Median	High
A2	2021–50	1.9	3.0	4.3
	2041–70	3.1	4.8	6.7
	2070–99	5.1	8.1	11.6
B1	2021–50	1.7	2.6	3.3
	2041–70	2.3	3.7	4.6
	2070–99	3.1	4.4	6.6

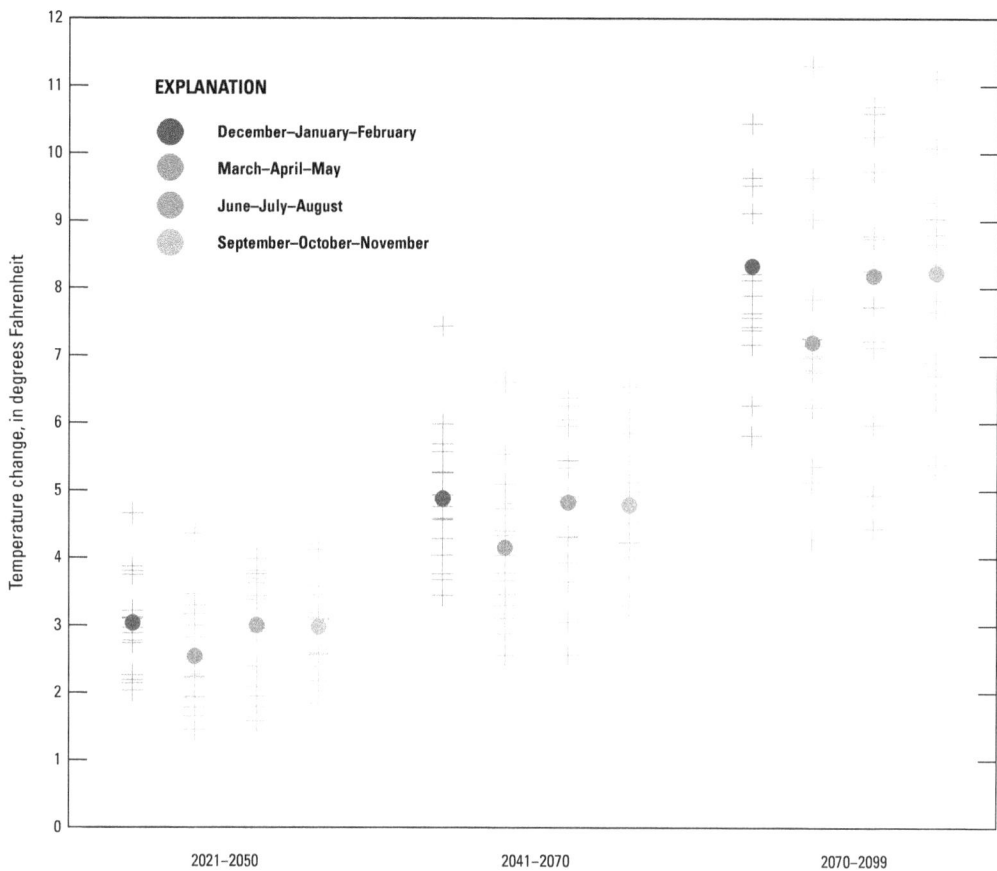

Figure 11. Mean seasonal temperature changes (°F) for each future time period with respect to the reference period of 1971–2000 for all 15 CMIP3 models, averaged over the entire Alaska region for the high (A2) emissions scenario. The plus signs are values for each individual model and the circles depict the overall means. CMIP3: Coupled Model Intercomparison Project Phase 3.

The preceding results have all been at the coarse resolution of the CMIP3 models. These same models have been used to produce downscaled output for Alaska at 2 km (1.2 mi) resolution. The downscaling has been carried out by the Scenarios Network for Alaska Planning (SNAP; University of Alaska Fairbanks, 2012a). The products are based on a subset of five CMIP3 models (table 4) that were found to best simulate the seasonal cycles of temperature, precipitation, and sea-level pressure over Alaska (Walsh and others, 2008). Temperature and precipitation products for the same forcing scenarios (A2, B1) are based on the Delta method and the Parameter-elevation Regressions on Independent Slopes Model climatology (PRISM; Oregon State University, 2012).

Figure 12 shows the downscaled [2 km (1.2 mi) resolution] annual mean temperature for 2000–2009 (top panel) and the corresponding downscaled projections in the B1 and A2 scenarios for 2060–2069 (corresponding to the middle time-slice in fig. 10). The prominent role played by topography is apparent in southeast Alaska as well as in the areas of the Alaska Range (south-central) and the Brooks Range, which separates the coastal plain to the north from the Alaskan interior to the south. Statewide warming generally increases from south to north and is greater in the A2 scenario than in the B1 scenario. Even in the warmer climate, however, spatial differences over scales of tens of miles are much larger than the changes resulting from external forcing.

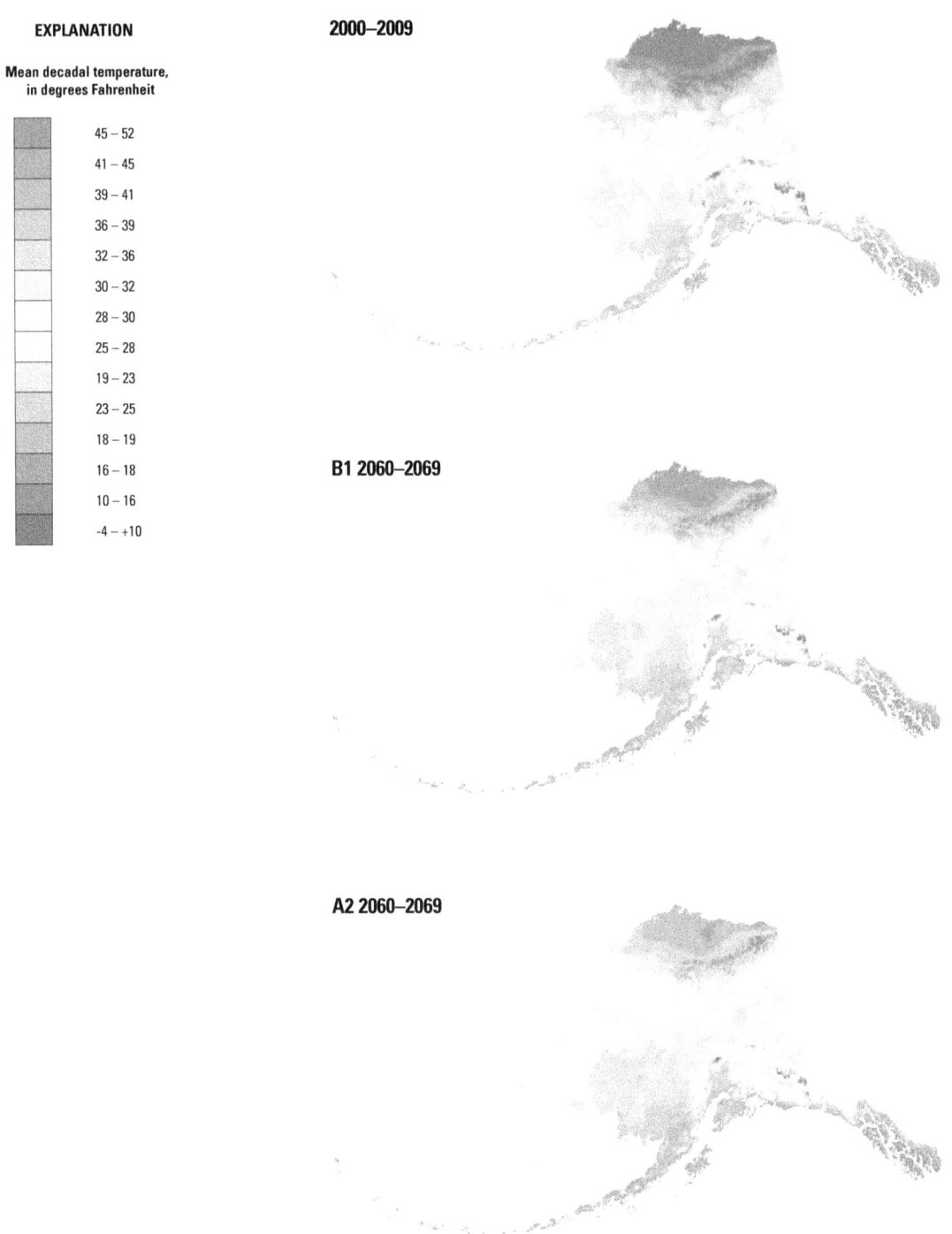

Figure 12. Downscaled fields of annual mean temperature (°F) from observational data for 2000–2009 (upper panel), CMIP3 model projections for 2060–2069 from B1 (middle panel), and A2 (lower panel) emissions scenarios. Downscaled projections are means from five CMIP3 models. Downscaled models produced by Scenarios Network for Alaska and Arctic Planning (University of Alaska Fairbanks, 2012a). CMIP3: Coupled Model Intercomparison Project Phase 3.

Table 4. Coupled Model Intercomparison Project Phase 3 models used for downscaling.

Model	Institution	Country	Year published	Publication
MPI ECHAM5	Max Planck Institute for Meteorology	Germany	2005	Jungclaus and others, 2005
GFDL CM2.1	Geophysical Fluid Dynamics Laboratory	United States	2005	Delworth and others, 2006; Gnanadesikan and others, 2006
MIROC3.2 (medres)	Center for Climate System Research, University of Tokyo	Japan	2004	Hasumi and Emori, 2004
UKMO HADCM3	Hadley Center for Climate Prediction and Research	United Kingdom	2000	Gordon and others, 2000 Pope and others, 2000
CCCMA CGCM3.1	Canadian Center for Climate Modeling and Analysis	Canada	2001	Flato and Boer, 2001 Flato and Hilber, 1992 Kim and others, 2002 Kim and others, 2003

The effect of each emission scenario (A2 and B1) on the seasonal temperature changes is depicted in figures 13 and 14, which show the projected temperature fields for winter and summer simulations, respectively, of the 2090–2099 time-slice. The differences in the two scenarios are especially apparent in the geographic extent of winter temperatures (fig. 13) on the North Slope. The area with temperatures of about -6°F in the A2 scenario has a much smaller footprint than that shown in the B1 scenario. The area with temperatures close to freezing during winter also is noticeably larger in the A2 scenario. In summer (fig. 14), the area with temperatures exceeding 57°F is larger in the A2 scenario. In both seasons, higher elevation areas are noticeably colder than low elevation areas, highlighting the need for downscaling when global model output is used in climate impact assessments for Alaska.

As specific examples of the downscaled output for Alaska, figures 15–17 show the seasonal cycles of the downscaled temperature for one historical reference period (1961–1990) for different time-slices of the B1 and A2 simulations: 2010–2019, 2040–2049, 2060–2069, and 2090–2099. The seasonal cycles are shown for three climatically different locations: (1) Anchorage (fig. 15)—a major population center with a generally maritime climate in the southern part of Alaska; (2) Fort Yukon (fig. 16)—a small village with a strongly continental climate in interior Alaska; and (3) Barrow (fig. 17)—a village on Alaska's northern coast, where the Arctic Ocean's sea ice impinges upon the coast for about 9 months of the year (although the length of the open water season is increasing).

Notable features of the temperature change at Anchorage are the pervasiveness of the warming (all calendar months) in both scenarios except for occasional single-decade decreases arising from natural variability; relatively small increases in spring (April–June) relative to those of the other seasons; the acceleration of the fall-winter warming in both scenarios; and the greater warming in the A2 scenario relative to B1 throughout the year. In comparison with Anchorage, Fort Yukon's projected changes show a similar seasonality and a similar dependence on the forcing scenario. However, Fort Yukon's projected warming is larger (consistent with fig. 9), especially in the winter. Barrow shows an even greater warming, in excess of 20°F during October–February by the end of the century in the A2 scenario.

EXPLANATION

B1 2090–2099

Mean winter temperature,
in degrees Fahrenheit

	32 – 45
	27 – 32
	19 – 27
	14 – 19
	9 – 14
	5 – 9
	-6 – +5

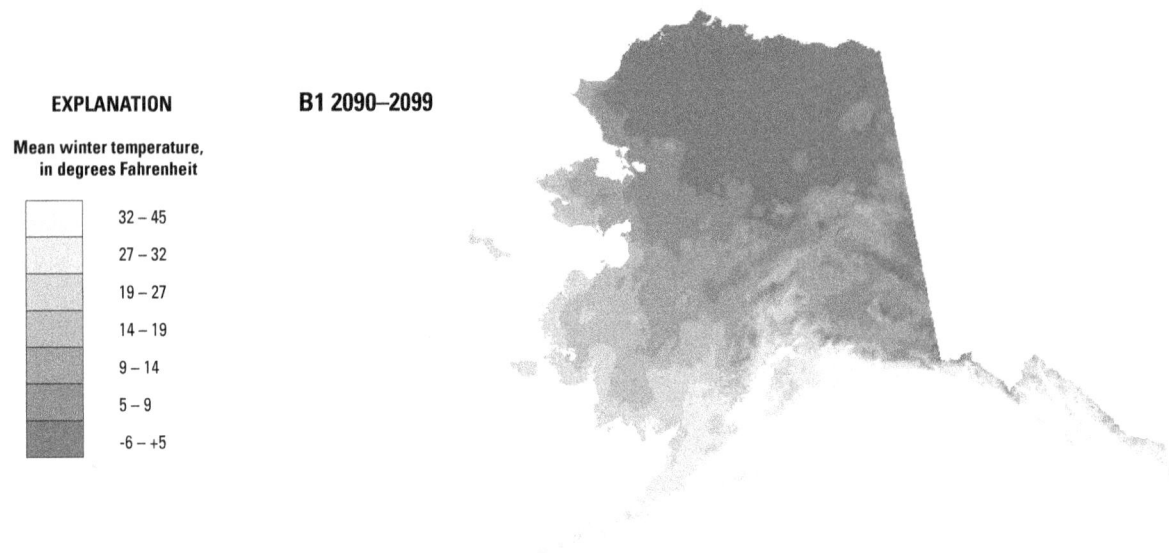

A2 2090–2099

Figure 13. Projected temperatures (°F) for winter (December–February) of the 2090–2099 decade in the B1 (upper panel) and A2 (lower panel) scenarios. Fields are composites of simulations from five different CMIP3 models (see table 4). Downscaled models produced by Scenarios Network for Alaska and Arctic Planning (University of Alaska Fairbanks, 2012a). CMIP3: Coupled Model Intercomparison Project Phase 3.

EXPLANATION

B1 2090–2099

Mean summer temperature,
in degrees Fahrenheit

61 – 68

59 – 61

57 – 59

55 – 57

54 – 55

50 – 54

7 – 50

A2 2090–2099

Figure 14.   Projected temperatures (°F) for summer (June–August) of the 2090–2099 decade in the B1 (upper panel) and A2 (lower panel) scenarios. Fields are composites of simulations from five different CMIP3 models (see table 4). Downscaled models produced by Scenarios Network for Alaska and Arctic Planning (University of Alaska Fairbanks, 2012a). CMIP3: Coupled Model Intercomparison Project Phase 3.

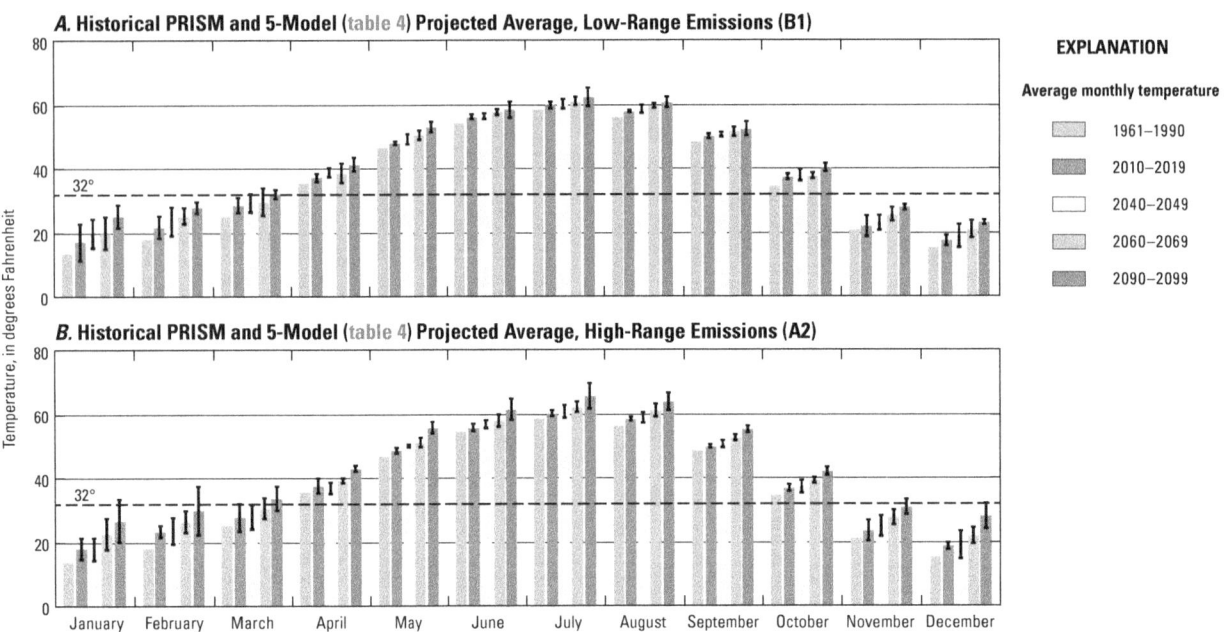

Figure 15. Projections of decadal mean temperatures (°F) by calendar month for Anchorage under B1 (upper panel) and A2 (lower panel) emissions scenarios. Across-model spread is indicated by thin black lines at tops of bars. Downscaled models produced by Scenarios Network for Alaska and Arctic Planning (University of Alaska Fairbanks, 2012a). PRISM: Parameter-elevation Regressions on Independent Slopes Model.

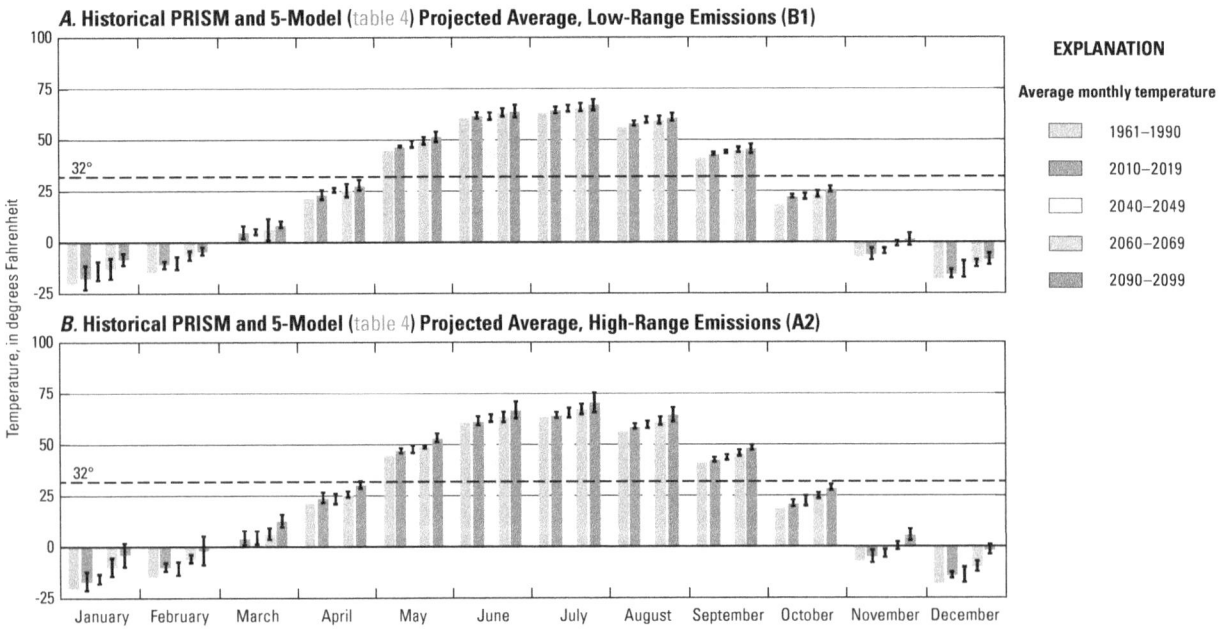

Figure 16. Projections of decadal mean temperatures (°F) by calendar month for Fort Yukon under B1 (upper panel) and A2 (lower panel) emissions scenarios. Across-model spread is indicated by thin black lines at tops of bars. Downscaled models produced by Scenarios Network for Alaska and Arctic Planning (University of Alaska Fairbanks, 2012a). PRISM: Parameter-elevation Regressions on Independent Slopes Model.

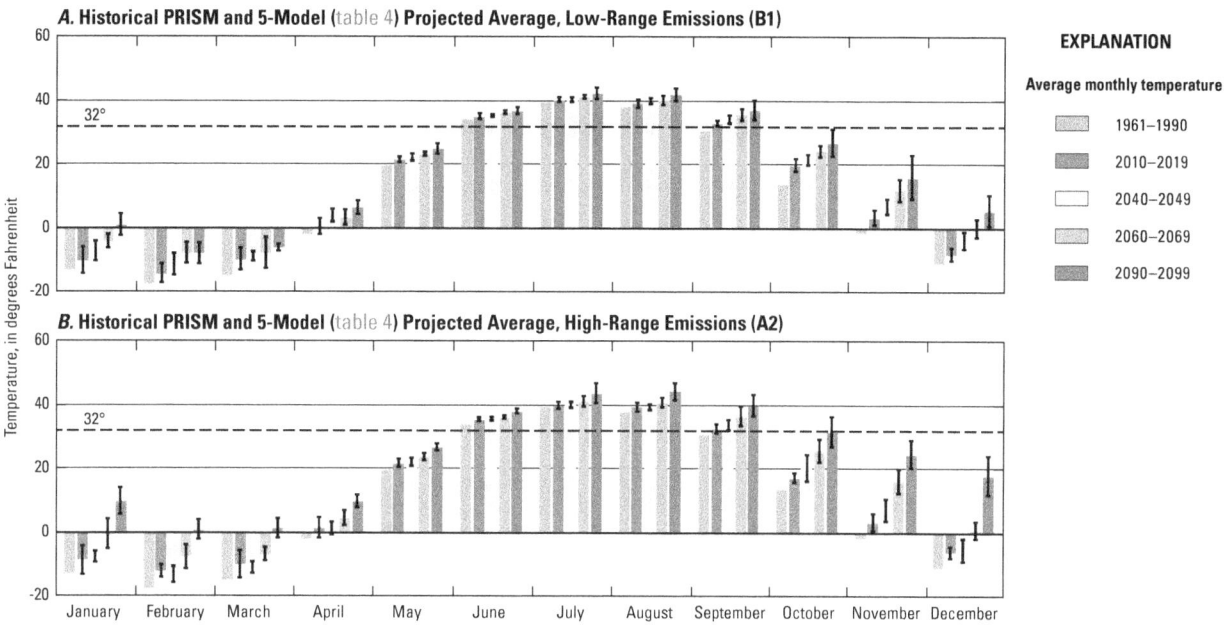

Figure 17. Projections of decadal mean temperatures (°F) by calendar month for Barrow under B1 (upper panel) and A2 (lower panel) emissions scenarios. Across-model spread is indicated by thin black lines at tops of bars. Downscaled models produced by Scenarios Network for Alaska and Arctic Planning (University of Alaska Fairbanks, 2012a). PRISM: Parameter-elevation Regressions on Independent Slopes Model.

The seasonality of the warming is similar in the B1 scenario for Barrow (fig. 17), although the warming in November–February period is only about 15°F by 2090–2099 in the B1 scenario. This large fall-winter warming is partly a consequence of the loss of sea ice in the climate models, as the longer duration of open water during spring and summer allows greater oceanic storage of heat that is subsequently released to the atmosphere in fall and early winter.

Finally, figure 18 shows the projected change in growing-season length, which is defined here as the number of days between the final freeze during spring and the first freeze of fall (for both the occurrence of a temperature of 32°F). These changes, shown in figures 19 and 20, are based on the daily output of the selected five CMIP3 models (table 4). Figure 18

shows the projected growing season lengths for 2060–2069, together with the corresponding distribution for 2000–2009, all downscaled to the same 2 km (1.2 mi) resolution of the temperature fields in figures 12–14. Increases of 2–3 weeks (15–25 days) are apparent in the southwestern and south-central parts of the State. In a large portion of southwestern Alaska, the growing season lengthens to more than 200 days, a value found only along the southern coastline and sub-Arctic islands in the present climate.

Figures 19 and 20 show that, by the last decade of the present century, the spring thaw date over much of interior Alaska is projected to advance by 2–3 weeks, while the fall freeze-up is delayed by about 2 weeks. The total change on the above-freezing period typically is about 30 days over much of the State.

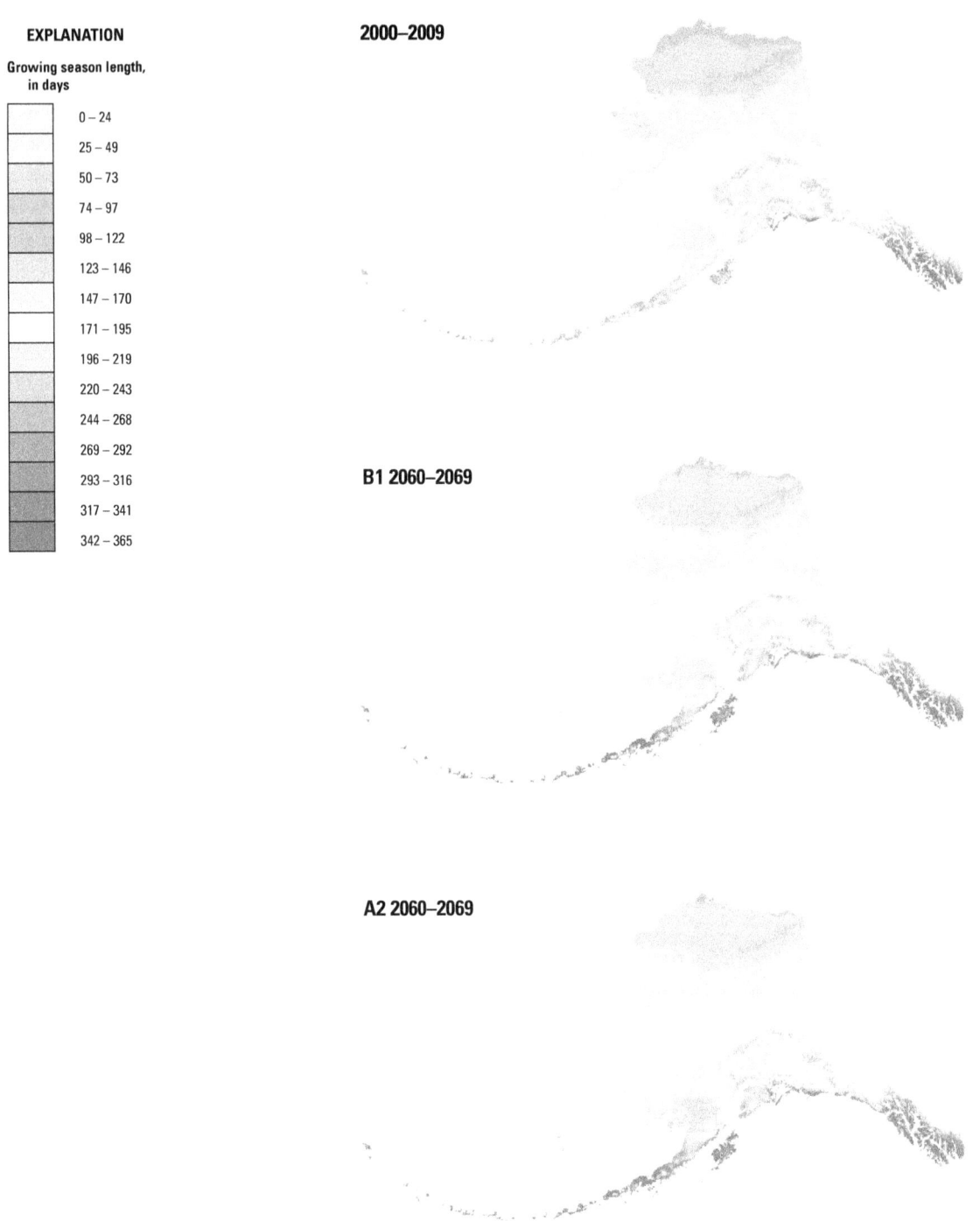

Figure 18. Downscaled distribution of length of growing season (days). Top panel shows observationally derived growing season lengths for 2000–2009. Projections are shown for 2060–2069 under B1 emission scenario (middle panel) and A2 emission scenario (lower panel). Downscaled models produced by Scenarios Network for Alaska and Arctic Planning (University of Alaska Fairbanks, 2012a).

EXPLANATION

Change in day of thaw (breakup)
1961–1990 to 2090–2099

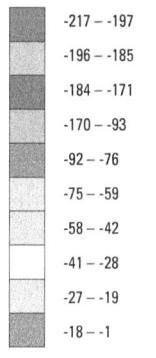

-217 – -197
-196 – -185
-184 – -171
-170 – -93
-92 – -76
-75 – -59
-58 – -42
-41 – -28
-27 – -19
-18 – -1

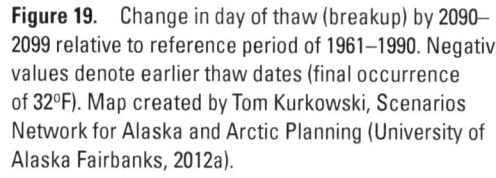

Figure 19. Change in day of thaw (breakup) by 2090–2099 relative to reference period of 1961–1990. Negative values denote earlier thaw dates (final occurrence of 32°F). Map created by Tom Kurkowski, Scenarios Network for Alaska and Arctic Planning (University of Alaska Fairbanks, 2012a).

EXPLANATION

Change in day of freeze (freezeup)
1961–1990 to 2090–2099

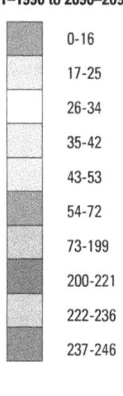

0-16
17-25
26-34
35-42
43-53
54-72
73-199
200-221
222-236
237-246

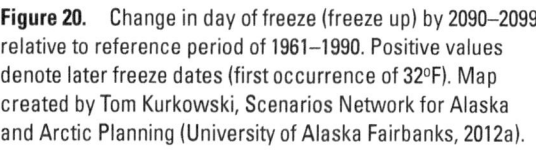

Figure 20. Change in day of freeze (freeze up) by 2090–2099 relative to reference period of 1961–1990. Positive values denote later freeze dates (first occurrence of 32°F). Map created by Tom Kurkowski, Scenarios Network for Alaska and Arctic Planning (University of Alaska Fairbanks, 2012a).

Precipitation Projections

The distribution of the CMIP3 multi-model mean changes in annual precipitation for three future periods (2021–2050, 2041–2070, and 2070–2099) and two emissions scenarios (A2 and B1) is shown in figure 21. An increase in precipitation can be seen in all cases, with the largest changes occurring in the far northwest of Alaska and the smallest changes in the southeast region. Spatial variations are again relatively small, with greater increases for the A2 scenario than for B1.

The gradient of changes also increases over time; for example, precipitation differences of between 0 and 15 percent are indicated for the A2 scenario in 2021–2050, whereas for 2070–2099, changes range from 10 to 35 percent.

The distribution of changes in mean annual precipitation for each future period and both emissions scenarios across the 15 CMIP3 models is shown in table 5. For all periods and both scenarios, the CMIP3 model simulations include both increases and decreases in precipitation. The median values are very small (2 to 4 percent) and the models range between -7 and 14 percent for the A2 scenario and -2 to 9 percent

for the B1 scenario. In both scenarios, the minimum and maximum range of values occurs in the 2070–2099 period. The inter-quartile range of changes across all the models is 9 percent or less.

Table 6 shows the seasonal distribution of precipitation changes across the 15 CMIP3 models, between 2070 and 2099 for both emissions scenarios. On a seasonal basis, the range of model-simulated changes is quite large. For example, in the A2 scenario, the change in summer precipitation ranges from a decrease of 23 percent to an increase of 16 percent. A majority of the models indicate increases in precipitation for all seasons, with the exception of summer for the A2 scenario. In the B1 scenario, the range of changes generally is smaller, with the largest decrease being 7 percent in summer. For both emissions scenarios, winter is the only season in which no models indicate a decrease in precipitation. The central feature of the results in table 6 is the large uncertainty in seasonal precipitation changes.

Figure 22 shows the mean annual changes in precipitation for each future time period and both emissions scenarios, averaged over the entire Alaska region for the individual 15 CMIP3 models. All mean changes are positive, although the values are small. For the A2 scenario, the CMIP3 models project changes in precipitation of between 2 and 5 percent, increasing for each future time period. For the B1 scenario, the values are comparable to those for A2 in 2021–2050 and 2041–2070, but lower for 2070–2099. The range of individual model changes in figure 22 is large compared to the differences in the multi-model means, as also illustrated in table 6. In fact, for all three future periods and for both emissions scenarios, the individual model range is much larger than the differences in the CMIP3 multi-model means.

Figure 23 shows the mean seasonal changes in precipitation for each future time period for the A2 scenario, averaged over the entire Alaska region for the individual 15 CMIP3 models. The decreases are largest in summer, ranging between -1 and -2 percent. For the other three seasons, precipitation changes are positive and increase over time, the largest being +10 percent for winter in 2070–2099. As was the case for the annual totals in figure 22, the model ranges in figure 23 are large compared to the multi-model mean differences. This illustrates the large uncertainty in the precipitation estimates derived by using these simulations.

Table 5. Distribution of changes in mean annual precipitation (percent) for the Alaska region for the 15 CMIP3 models.

[A2, emissions scenario that assumes a continuation of recent trends in fossil fuel use. B1, emissions scenario that assumes a vigorous global effort to reduce fossil fuel use. CMIP3: Coupled Model Intercomparison Project Phase 3]

Scenario	Period	Low	25th percentile	Median	75th percentile	High
A2	2021–2050	-3	0	2	4	6
	2041–2070	-4	0	2	6	9
	2070–2099	-7	-1	4	9	14
B1	2021–2050	-2	1	2	3	5
	2041–2070	-1	0	3	5	7
	2070–2099	-2	2	4	6	9

Table 6. Distribution of changes in mean seasonal precipitation (percent) for the Alaska region for the 15 CMIP3 models.

[A2, emissions scenario that assumes a continuation of recent trends in fossil fuel use. B1, emissions scenario that assumes a vigorous global effort to reduce fossil fuel use. CMIP3: Coupled Model Intercomparison Project Phase 3]

Scenario	Period	Season	Low	Median	High
A2	2070–2099	December–January–February	3	10	18
		March–April–May	-9	5	15
		June–July–August	-23	-4	16
		September–October–November	-13	8	17
B1	2070–2099	December–January–February	1	5	11
		March–April–May	-6	5	10
		June–July–August	-7	1	13
		September–October–November	-5	5	12

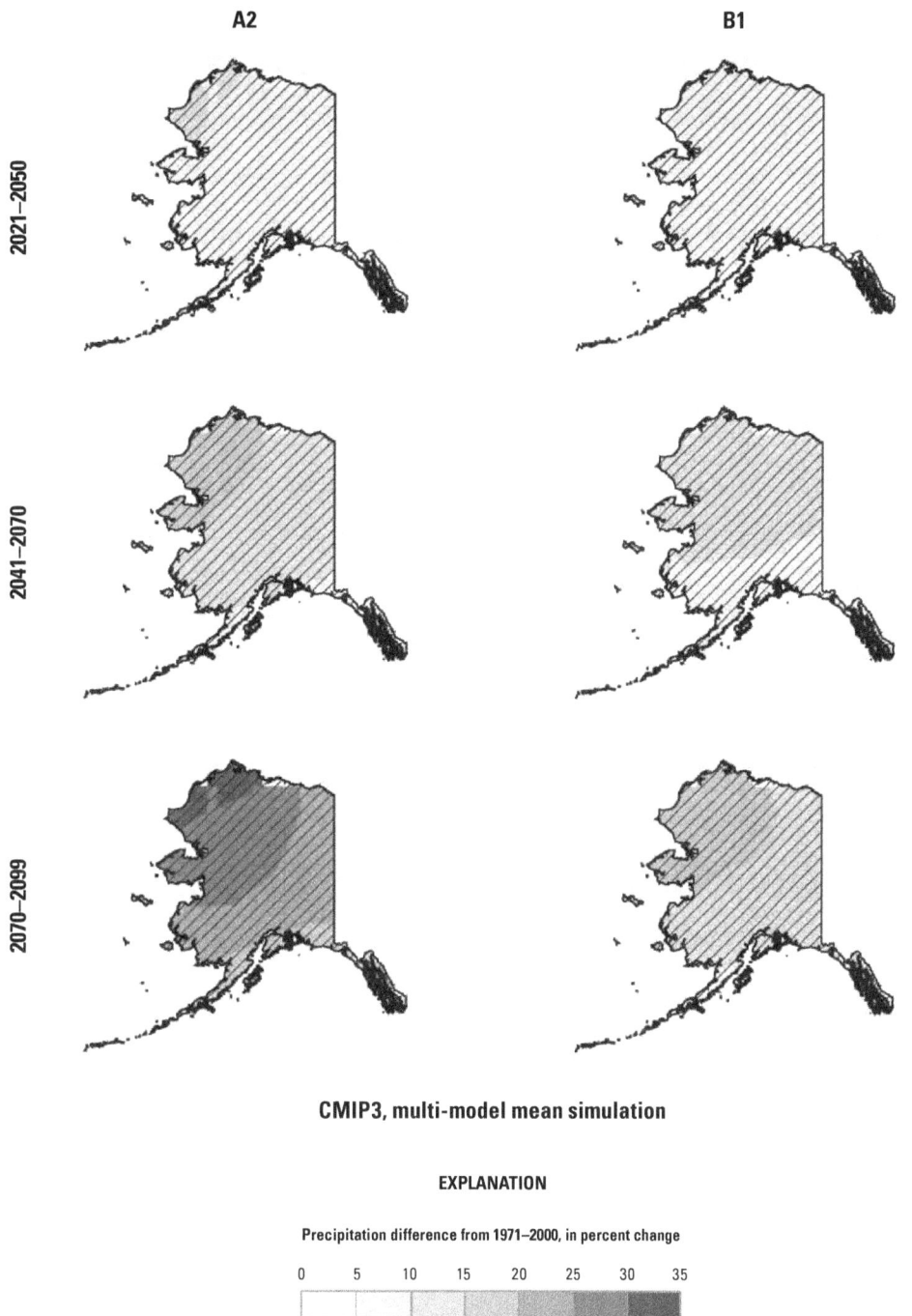

CMIP3, multi-model mean simulation

EXPLANATION

Precipitation difference from 1971–2000, in percent change

0   5   10   15   20   25   30   35

Figure 21.   Multi-model mean annual differences in precipitation (percent) between three future periods and 1971–2000, from the 15 CMIP3 model simulations. Areas with hatching indicate that more than 50 percent of the models show a statistically significant change in precipitation. Map created by Michael Wehner, Lawrence Berkeley National Laboratory, 2011. CMIP3: Coupled Model Intercomparison Project Phase 3.

Precipitation in Alaska shows a spatially complex pattern, which is largely a result of the State's major topographic features: the coastal mountains of the southeast, the Alaska Range in the south-central region, and the Brooks Range in the north. Figure 24 shows downscaled fields of annual total precipitation for the present-day climate (2000–2009) and for 2060–2069 under the B1 and A2 scenarios. As in the case of figure 19, the projections are composited over simulations by the five CMIP3 models that best capture the seasonal cycles of key climate variables in Alaska over the past several decades. All panels of figure 24 show a large range in annual precipitation, with less than 250 mm on the North Slope and more than several meters in the southeast. Both scenarios indicate general increases by the decade of the 2060s. The

actual increases (amounts) are largest in the southeast, although the percentage increases are largest in the western and northern parts of the State.

In order to illustrate the seasonality of the changes of precipitation locally, figures 25–27 show decadal time-slices of calendar-month mean precipitation for the same locations for which temperature projections were shown earlier (figs. 15–17): Anchorage near the southern coast (fig. 25), Fort Yukon in the interior (fig. 26), and Barrow on the northern coast (fig. 27). Several noteworthy features of the precipitation amounts and projected changes can be seen in figures 25–27. First, seasonal precipitation amounts are about twice as high at Anchorage than at the other two stations (note the different scales used for precipitation amounts). This difference however lessons under future projections for some months.

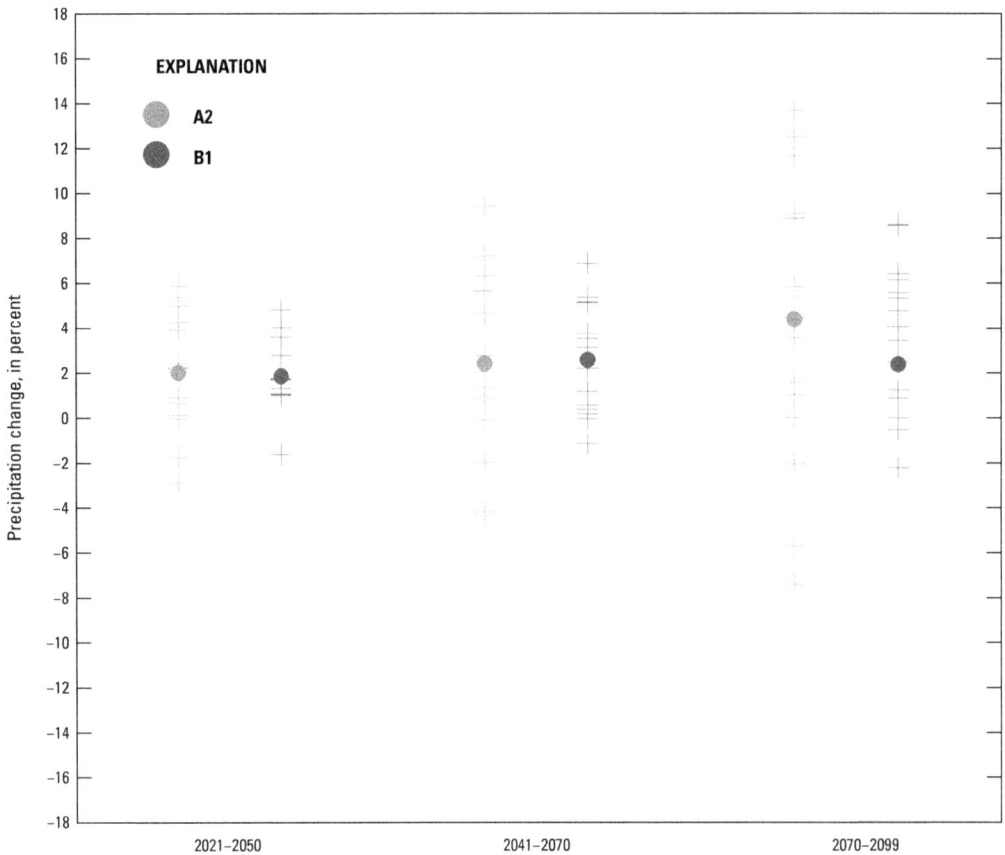

Figure 22. Mean annual precipitation changes (percent) for each future time period with respect to the reference period of 1971–2000 for all 15 CMIP3 models, averaged over the entire Alaska region for the high (A2) and low (B1) emissions scenarios. The plus signs are values for each individual model and the circles depict the overall means. CMIP3: Coupled Model Intercomparison Project Phase 3.

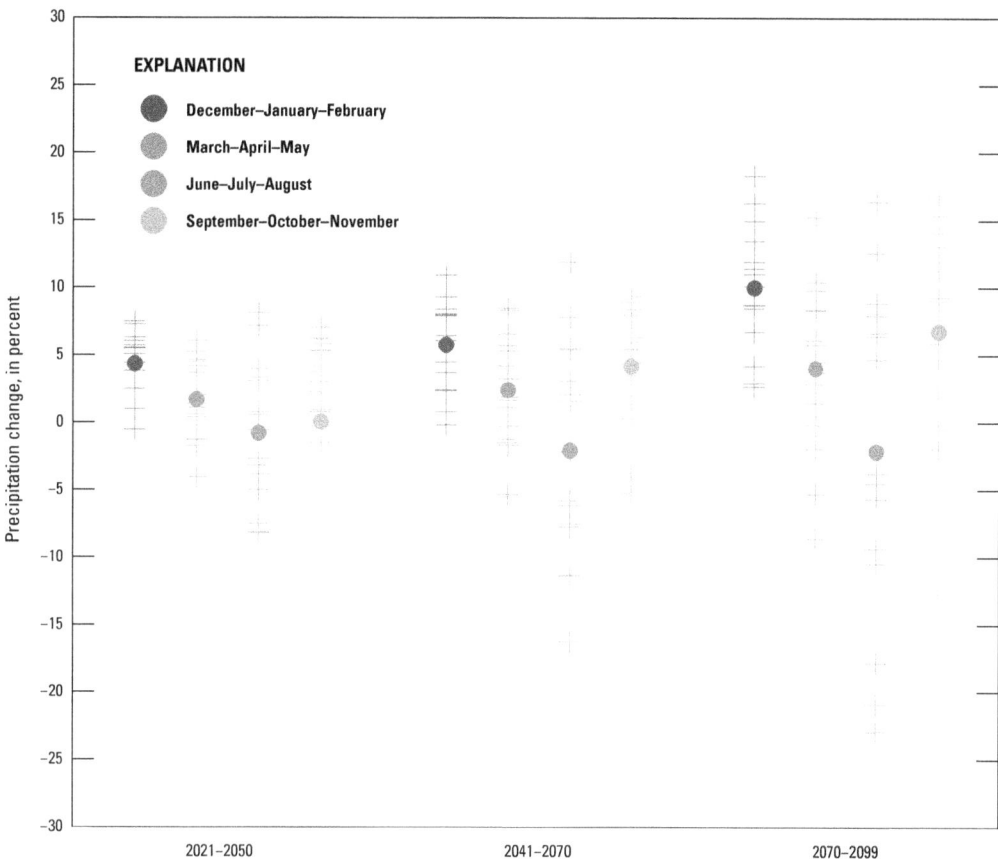

Figure 23. Mean seasonal precipitation changes (percent) for each future time period with respect to the reference period of 1971–2000 for all 15 CMIP3 models, averaged over the entire Alaska region for the high (A2) emissions scenario. The plus signs are values for each individual model and the circles depict the overall means. CMIP3: Coupled Model Intercomparison Project Phase 3.

Second, precipitation amounts generally increase in all calendar months at all stations in both scenarios, although each station shows a few examples of future decadal decreases due to natural variability. The fact that such decreases occur even though the bars in the figures represent five-model means points to the importance of natural variability on decadal timescales.

Third, the A2 scenario shows an acceleration of the increase of precipitation in the final decade, especially in summer and fall. The possible role of increased open water (loss of sea ice) in these large increases merits investigation, especially because a similar type of behavior was apparent in the corresponding temperature plots. The prominence of this acceleration in scenario A2 but not in scenario B1 results also merits closer examination, as it may point to an impact of mitigation if mitigation actions indeed shape future greenhouse gas concentrations.

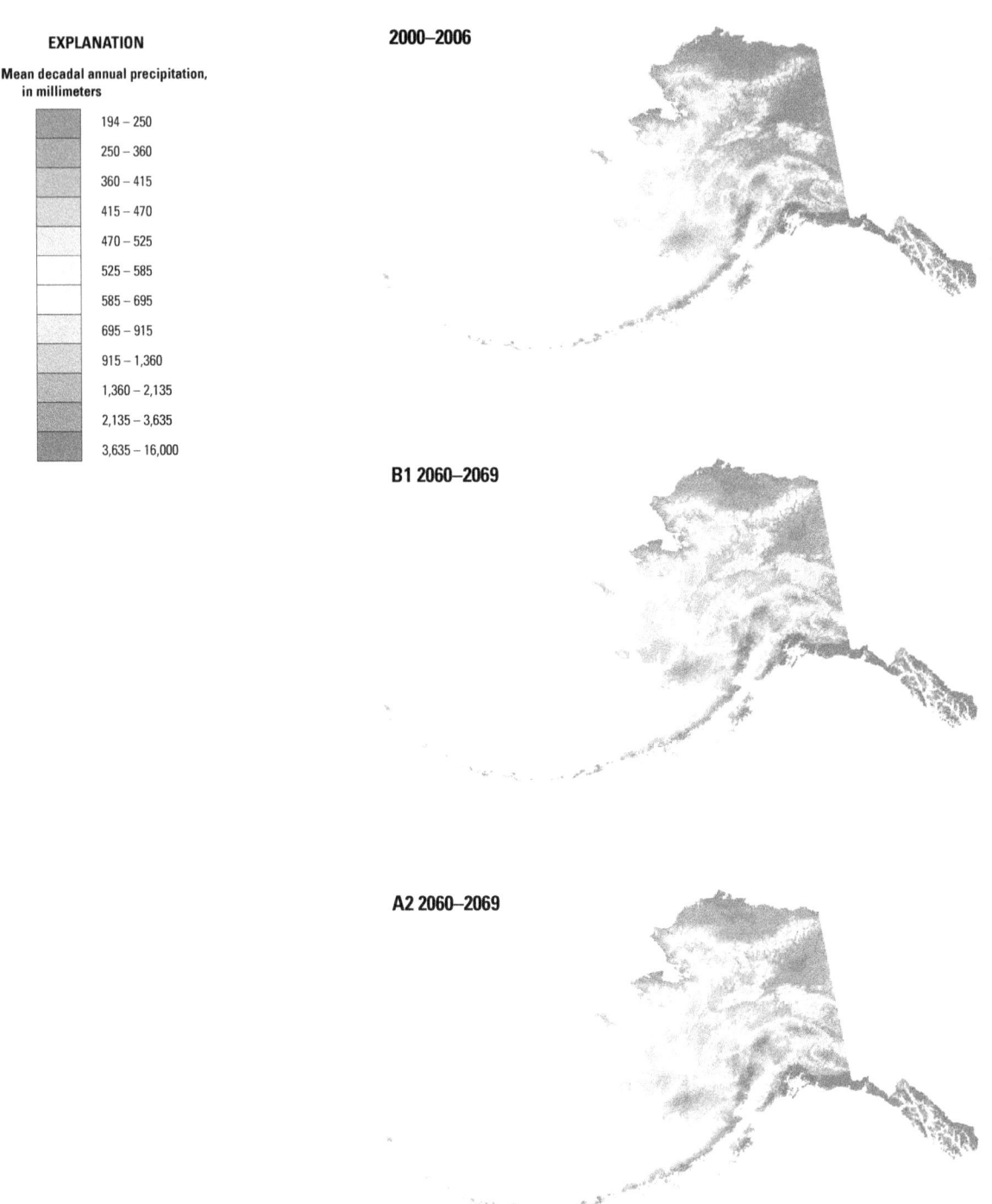

EXPLANATION

Mean decadal annual precipitation,
in millimeters

194 – 250
250 – 360
360 – 415
415 – 470
470 – 525
525 – 585
585 – 695
695 – 915
915 – 1,360
1,360 – 2,135
2,135 – 3,635
3,635 – 16,000

2000–2006

B1 2060–2069

A2 2060–2069

Figure 24. Downscaled fields of annual total precipitation: observationally derived for 2000–2006 (upper panel), projections for 2060–2069 from B1 (middle panel), and A2 (lower panel) emissions scenarios. Downscaled projections are means from five CMIP3 models (table 4). Downscaled models produced by Scenarios Network for Alaska and Arctic Planning (University of Alaska Fairbanks, 2012a). CMIP3: Coupled Model Intercomparison Project Phase 3.

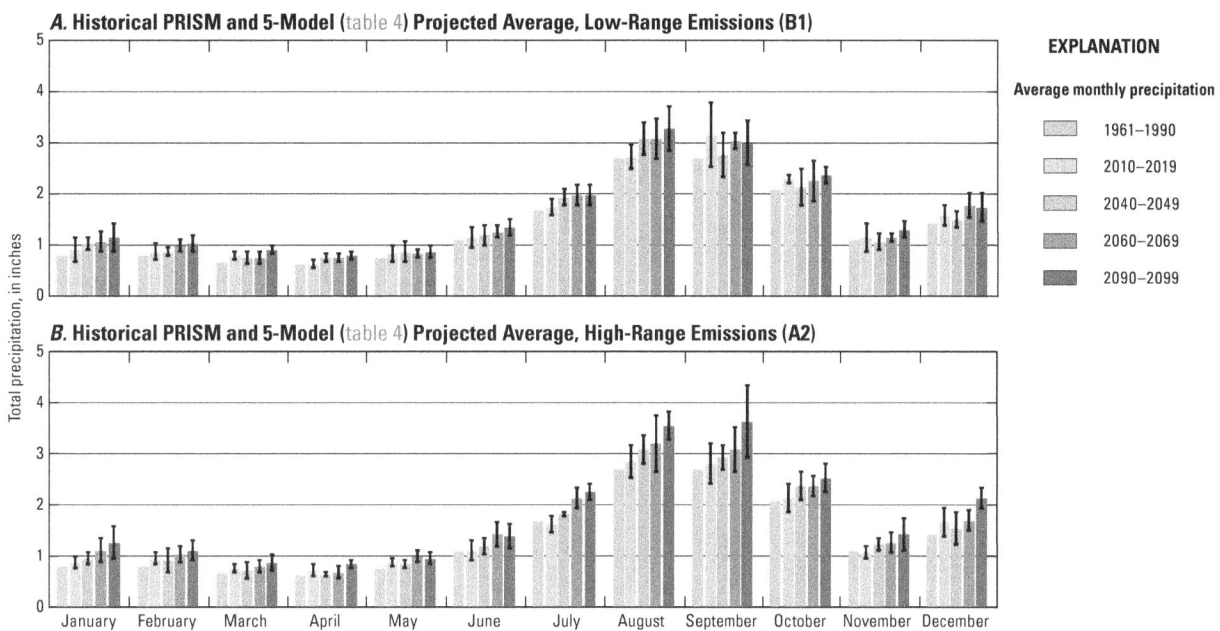

Figure 25. Downscaled projections of precipitation (inches) by calendar month for Anchorage under B1 (upper) and A2 (lower) emissions scenarios. Across-model spread is indicated by thin black lines at tops of bars. Downscaled models produced by Scenarios Network for Alaska and Arctic Planning (University of Alaska Fairbanks, 2012a). PRISM: Parameter-elevation Regressions on Independent Slopes Model.

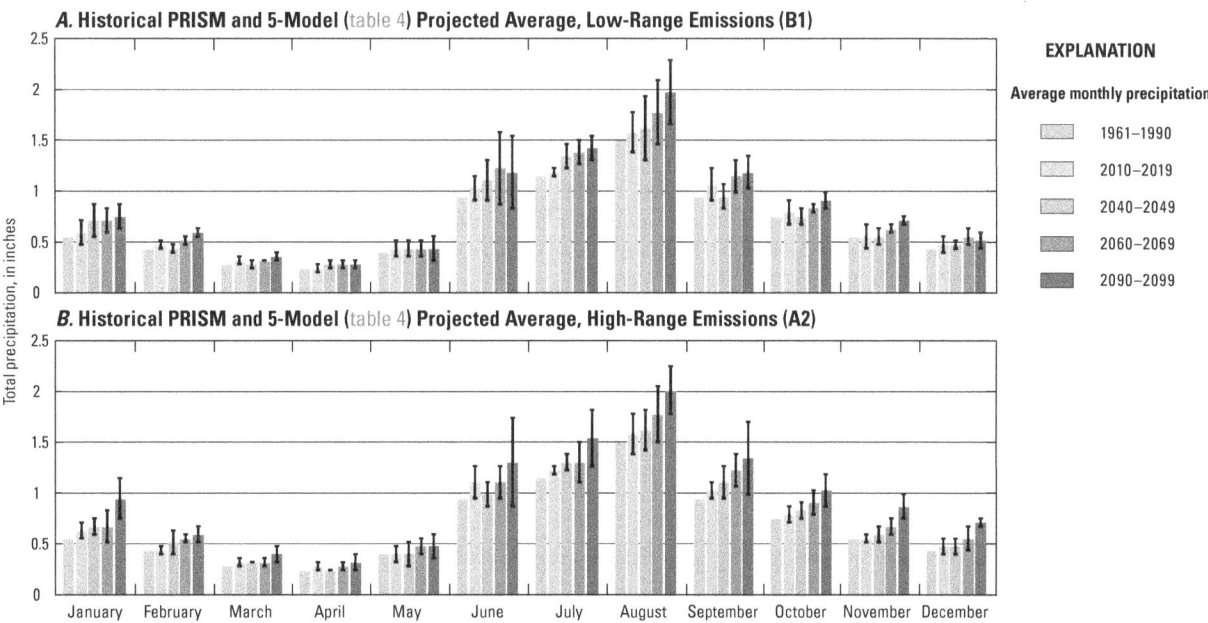

Figure 26. Downscaled projections of precipitation by calendar month for Fort Yukon under B1 (upper) and A2 (lower) emissions scenarios. Across-model spread is indicated by thin black lines at tops of bars. Downscaled models produced by Scenarios Network for Alaska and Arctic Planning (University of Alaska Fairbanks, 2012a). PRISM: Parameter-elevation Regressions on Independent Slopes Model.

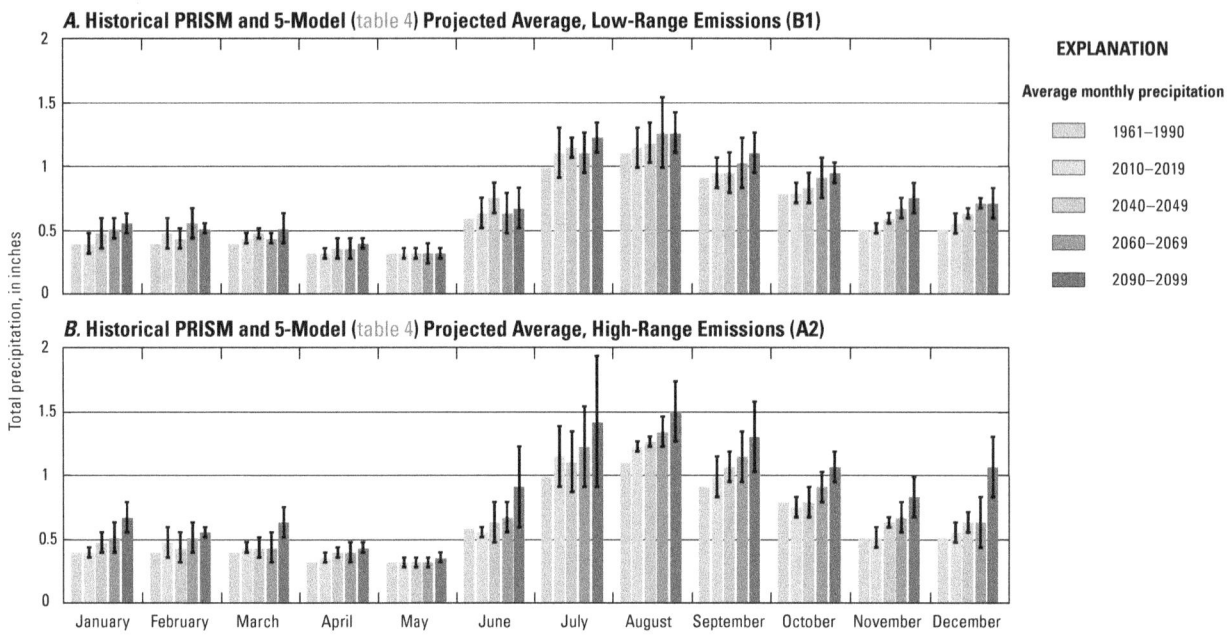

Figure 27. Downscaled projections of precipitation by calendar month for Barrow under B1 (upper) and A2 (lower) emissions scenarios. Across-model spread is indicated by thin black lines at tops of bars. Downscaled models produced by Scenarios Network for Alaska and Arctic Planning (University of Alaska Fairbanks, 2012a). PRISM: Parameter-elevation Regressions on Independent Slopes Model.

Permafrost Projections

Permafrost depths and temperatures are determined primarily by two climate variables—mean annual temperature and wintertime snow depth—and by soil properties and slope (aspect) of the ground. Because snow effectively insulates the ground during winter, the boundary of permafrost typically is near the -1 to -2°C isotherm of mean annual temperature.

A climate model output has been used to drive permafrost models in order to capture the degradation of permafrost under greenhouse-driven climate projections based on the University of Alaska's Geophysical Institute Permafrost Laboratory model, using downscaled (2 km/1.2 mi) temperature and precipitation (snowfall) output from the B1 and A2 simulations and the five models described in table 4.

Figure 28 shows the 2-km (1.2-mi) resolution annual mean ground temperatures at 1-m depth from the B1 and A2 simulations for three time-slices: 2000–2009, 2040–2049, and 2090–2999. An annual mean ground temperature of 0°C at 1-m depth is an effective indicator of stable permafrost (T < 0°C) or degrading/non-permafrost (T > 0°C). In the figures below, the 1-m depth temperatures are color-coded in blue for temperatures below 0°C and red for temperatures above 0°C. It is apparent from the figures that the area of above-freezing ground expands dramatically during the 21st century under both scenarios, especially in the second half of the century. In the A2 scenario, the projected area of permafrost degradation (change from blue to red) by the last decade of the century covers most of interior Alaska. The thaw is more discontinuous in the B1 simulation, although patches of thaw extend as far north as the Brooks Range even under the B1 scenario.

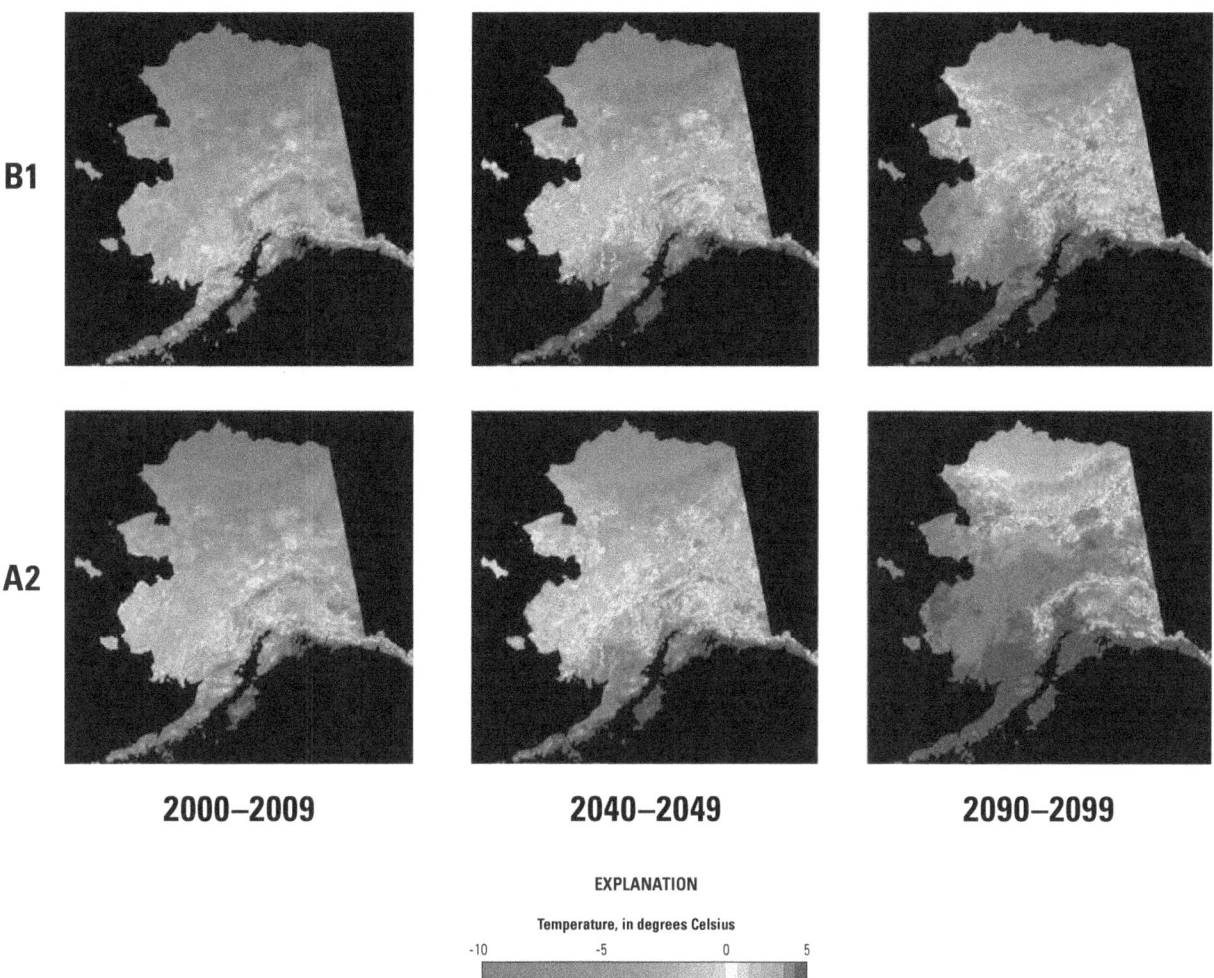

B1

A2

2000–2009   2040–2049   2090–2099

EXPLANATION

Temperature, in degrees Celsius

-10   -5   0   5

Figure 28. Mean annual ground temperatures at 1-m depth in permafrost model simulations driven by output from climate models run under B1 (upper panels) and A2 (lower panels) emissions scenarios. As indicated by color bars, blue shades represent temperatures below 0°C; red shades represent temperatures above 0°C. Figure provided by S. Marchenko and the Geophysical Institute Permafrost Laboratory, University of Alaska, Fairbanks.

Sea Ice Projections

On a regional basis, climate models project large declines in sea ice extent in the Alaskan region. Figure 29 shows the projections from CMIP3 models that were found to produce the most realistic simulations of recent sea ice conditions in the Alaskan subregions. Summer sea ice in the Chukchi Sea disappears between 2030 and 2050 in some models, while winter sea ice in the Bering and Chukchi Seas decrease by more than 50 percent by the end of the century (Overland and Wang, 2007). However, these may underestimate the decline of sea ice because sea ice appears to be declining in a non-linear manner (Stroeve and others, 2011). Wang and Overland (2009) predicted a nearly sea ice free Arctic summer by 2037, while others have suggested this could occur sooner (Maslowski and others, 2008; Zhang, 2010). It should be noted that the simulations summarized in figure 29 are based on the A1B scenario[3]; the rates of ice loss are larger (smaller) in the A2 (B1) scenario.

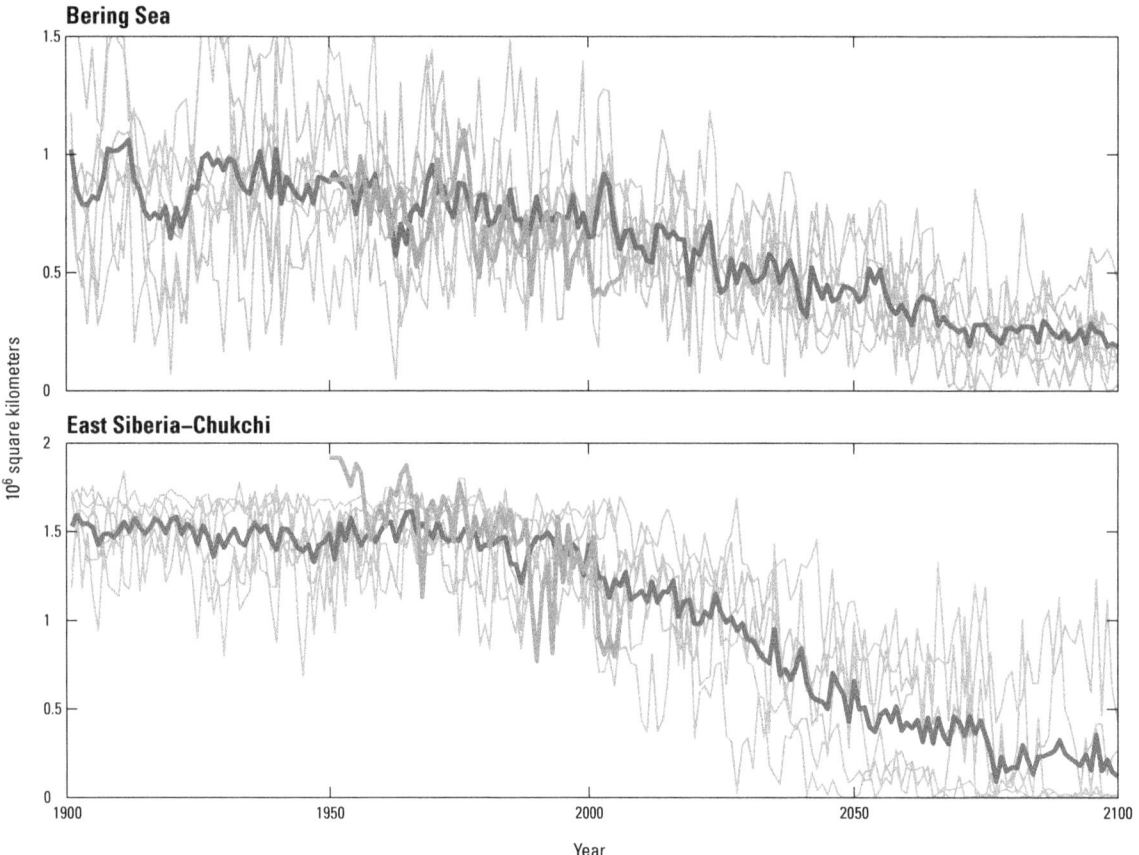

Figure 29. Winter (March–April average) sea ice area in the Bering Sea and summer (August–September average) sea ice area in the East Siberia-Chukchi Seas as projected by different CMIP3 models (thin grey lines). The thick blue line is multi-model mean, and the thick red line shows recent observational data (modified from Overland and Wang, 2007). CMIP3: Coupled Model Intercomparison Project Phase 3.

---

[3]The A1B scenario assumes a balanced use of resources not relying to heavily on one particular energy source. It provides a good mid-line scenario for carbon dioxide output and economic growth.

Observed Environmental Trends

Numerous changes have taken place across the Alaskan landscape and surrounding ocean environment, many of which can be related to climate change. To address these changes, this section of the report is divided into four major areas—ocean environment, hydrologic linkages, land environment, and human environment. Some of the subsections below provide information on topics that have been reported in common media outlets (such as reduction of Arctic sea ice) and are well known. Other areas of interest or concern, however, are not as widely recognized, such as the effects of climate change on human health or warming permafrost. Although this synthesis is by no means exhaustive, it summarizes the major processes and activities that are affected by a changing climate in Alaska.

The Ocean Environment

Alaska's continental shelf seas, the Gulf of Alaska and the Bering, Chukchi, and Beaufort Seas, are linked to one another via a sea circulation consisting of a counterclockwise flow around the Gulf of Alaska (Spies and Weingartner, 2007) of which a portion crosses over to the Bering Sea shelf (Weingartner and others, 2005; Aagaard and others, 2006). Bering shelf waters incorporate waters from the Bering Sea basin as they flow northward through Bering Strait (Coachman and others, 1975; Woodgate and others, 2006). These Pacific waters thence flow across the Chukchi shelf (Weingartner and others, 2005; Woodgate and others, 2005) and around the western Arctic Ocean including the Beaufort Sea shelf, and eventually into the North Atlantic Ocean (Aagaard and Carmack, 1989). From a global perspective, this transport represents an important component of the global hydrologic cycle, for it returns the relatively fresh waters of the northern North Pacific to the saltier North Atlantic Ocean (Aagaard and Carmack, 1989; Wijffels and others, 1992). This transport also maintains the stratification and ice cover of the Arctic Ocean (Aagaard and others, 1981) and it exerts an important influence on the global thermohaline circulation (Wadley and Bigg, 2002). Regionally, this transport connects the marine ecosystems of Alaska's shelf seas and thus represents a pathway by which organisms, nutrients, heat, and freshwater are transferred from one shelf to another. Moreover, physical and biogeochemical processes operating over each shelf modify the waters along this pathway. Several of these processes are susceptible to regional climate change. In particular, the position and strength of the Aleutian Low governs the atmospheric properties that influence the circulation, mixing, and exchange of heat and fresh water between these seas and the atmosphere. Analysis of climate models by the Intergovernmental Panel on Climate Change Fourth Assessment Report (2007), suggest that the Aleutian Low may slightly intensify and translate northward under future warming scenarios (Salathé, 2006).

The Gulf of Alaska shelf circulation is controlled primarily by winds and the massive freshwater runoff entering the Gulf along the British Columbian and Alaskan coasts. On a mean annual basis, this runoff constitutes about 480 mi^3/yr (Royer, 1982; Weingartner and others, 2005; Neal and others, 2010; Morrison and others, 2011). The runoff controls shelf stratification and mixing, fronts, and the strength of the along-shelf flow. It also represents an influx of nitrate-poor, but iron- and sediment-rich waters that influence biological production. A warmer atmosphere, as predicted by the Intergovernmental Panel on Climate Change (2007), can hold more moisture and increase glacial melting (Arendt and others, 2002) and precipitation rates (Cassano and others, 2006), which would lead to increased annual runoff volume. The seasonal timing of coastal discharge also may be altered. Winter precipitation typically is bound in snowpacks and released from summer through autumn so that the discharge is maximal in summer during the melt season and in the autumn when coastal precipitation rates are greatest (Royer, 1982). Warmer winters are expected to increase winter runoff rates, leading to an earlier onset of ocean stratification and the spring bloom on the shelf (Janout and others, 2010). Because winter and early spring stratification affects the ocean temperature distribution throughout summer (Janout and others, 2010), these changes may influence a variety of ecosystem processes on this shelf.

In the early 1970s, the Bering Sea shelf experienced a period of extensive ice cover and low temperatures, which then gave way to a period of warmer temperatures and reduced ice extent (Niebauer, 1998; Hare and Mantua, 2000). Sea surface temperatures generally have warmed since the 1970s, with this rate of warming increasing since about 1990 (Steele and others, 2008). These changes were accompanied by an increase in the duration of ice-free waters not only in the Bering Sea (fig. 30; Stabeno and others, 2007; Danielson and others, 2011), but also in both the Chukchi and Beaufort Seas. As the climate warms, the duration of the late summer ice-free season will increase, although sea ice is expected to form during the winter on all of these shelves. Because many ecosystem processes on the Bering Sea shelf depend on sea-ice processes (Hunt and others, 2002; Grebmeier and others, 2006), reductions in the duration of the ice-free season and the extent of the ice cover might lead to changes in this ecosystem.

Glacial-Flour-Derived Dust as an Important Source of Iron to the Gulf of Alaska

The waters of the Gulf of Alaska are highly productive and support numerous economically important fisheries (Denman and others, 1981). There is evidence that the long-term average fish production along the continental margin of western North America, including in the northern Gulf of Alaska, is controlled by phytoplankton production (Ware and Thomson, 2005). It has become clear that the micronutrient iron limits phytoplankton productivity in much of the Gulf of Alaska and the sub-Arctic North Pacific (Boyd and others, 2004). Yet there is a poor quantitative understanding of the sources of iron, which include rivers, coastal eddies, dust, upwelling, and sediment remobilization from the continental shelf. Furthermore, we know that glaciers in the region are melting rapidly (Arendt and others, 2002). This meltwater is an important source of both dissolved and particulate iron, derived from glacial flour (finely ground glacial sediment), and of fresh water to the Gulf of Alaska. However, the impacts of climate change on these interconnected processes are poorly known.

Recent work has examined two sources of iron and examined possible climate change impacts. Schroth and others (2011) examined the concentrations and chemical form of iron in a series of tributaries of the Copper River. Iron fluxes from glacierized tributaries maintain high suspended sediment and colloidal iron loads of mixed valence silicate species, with low concentrations of dissolved iron and dissolved organic carbon (DOC). Conversely, iron fluxes from boreal-forested systems have higher concentrations of dissolved iron owing to higher DOC concentrations. The work by Schroth and others (2011) predicts that as the Gulf of Alaska watershed evolves due to deglaciation, so will the source, flux, and chemical nature of riverine iron loads, which could have significant ramifications for Alaskan marine and freshwater ecosystems.

Work by Crusius and others (2011) provided the first description of widespread glacial-flour-derived dust transport from coastal Alaska and suggested that this is an important source of iron to the Gulf of Alaska. Dust is frequently transported from glacially derived sediment at the mouths of several rivers, the most prominent of which is the Copper River. These dust events occur most frequently in fall, when coastal river levels are low and riverbed sediments are exposed. The dust plumes are transported several hundred kilometers beyond the continental shelf into iron-limited waters. The vast majority of glaciers that drain into the Gulf of Alaska are currently (2012) both retreating and losing mass (Arendt and others, 2002), which might suggest that dust fluxes are increasing in response, although there are currently (2012) no data to confirm this suggestion. Time-series air sampling has begun on Middleton Island (see figure below) that may help test this hypothesis. Additional work will be need to further clarify the impacts of these and other changing iron fluxes on the Gulf of Alaska ecosystem.

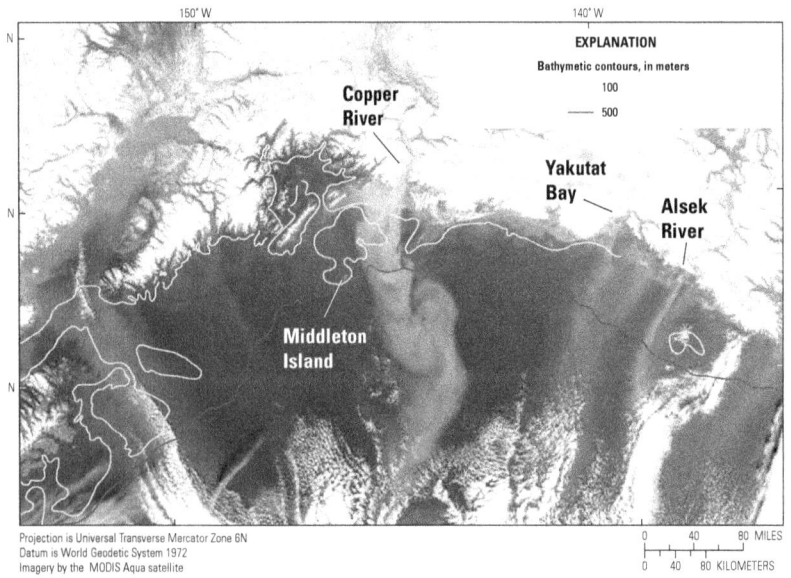

A glacial flour dust storm emanating from riverbed sediments of coastal Alaska on November 6, 2006 (from Crusius and others, 2011). The dust plume extends hundreds of kilometers beyond the continental shelf break (roughly represented by the 500-m contour line, in blue), into waters where iron is thought to be the nutrient limiting biological productivity. These events occur at least annually. The 100–m bathymetric contour also is shown (white line). True-color images were made from Moderate Resolution Imaging Spectro-radiometer (MODIS) data using hdflook (http://www-loa.univ-lille1.fr/Hdflook/hdflook_gb.html).

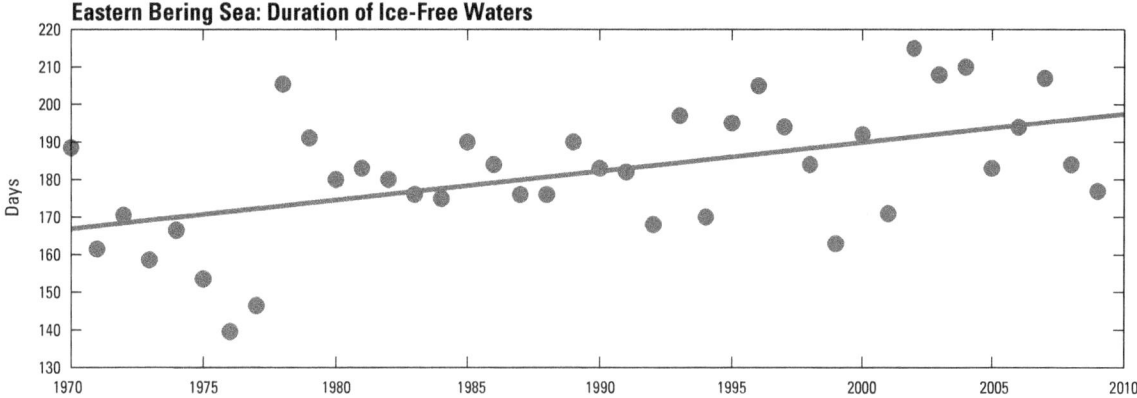

Figure 30. Duration of ice-free waters over the eastern Bering Sea shelf for the period 1970–2009 (from Danielson and others, 2011).

Climate-induced changes imparted to the Bering Sea will be advected into the Chukchi Sea through the Bering Strait. It seems likely that both the heat and freshwater flux will increase through the strait (Woodgate and others, 2006, 2010). This oceanic heat flux plays an important role in the seasonal sea ice cycle of the Chukchi shelf (Woodgate and others, 2005, 2010); it initiates earlier than otherwise expected ice retreat in spring and delays ice formation in fall. An increase in freshwater flux (manifested as a decrease in the salinity of the water flowing through Bering Strait) should increase stratification on the Chukchi shelf and may signify a decrease in the nutrient load, because low-salinity waters tend to have lower nutrient concentrations (Walsh and others, 1989). Changes in the seasonal sea ice distribution, the oceanic heat flux, and nutrient loads will likely lead to changes in biological production on this shelf. For example, warmer shelf waters may provide more optimal thermal habitats for some species, a reduction in ice cover may increase pelagic phytoplankton production, and reduced ice habitats may affect the distribution of some marine mammals.

Changes in sea-ice distribution (thickness and concentration), prolongation of the ice-free season, and freshwater discharge also will affect the circulation on the Beaufort Sea. For example, approximately 25 percent of the Alaskan Beaufort shelf is covered by landfast ice between mid-October and late June. This ice occupies the inner shelf and, because it is immobile, inhibits the wind-forced circulation here. Landfast ice effectively creates two different shelf circulation regimes. Seaward of the landfast ice, the sea-ice and ocean are energetically propelled by and directly respond to the wind. Within the landfast ice zone, the under-ice circulation is weak and the wind-current response is

indirect and uncorrelated (Weingartner and others, 2009; Kasper and Weingartner, 2012). The different ice regimes also result in decreased exchange between the inner and outer portions of the Beaufort shelf. The landfast ice also protects the inner shelf and coast from pack ice forces (Barnes and others, 1984), and, in fall, protects the coast from storm waves. As landfast ice thickness diminishes, the shelf area occupied by the landfast ice will decrease, changing patterns of ice-gouging on the inner shelf. Atmospheric and oceanic warming will delay the onset of landfast ice formation in fall when storm winds and waves generally are greatest and lead to increased sediment re-suspension and transport on the inner shelf and to increased rates of coastal erosion.

A diminished ice pack will increase regional atmospheric warming and precipitation (Rawlins and others, 2010; Stroeve and others, 2011). Increased precipitation and terrestrial permafrost melting will increase coastal erosion and alter the terrestrial hydrologic cycle of Alaska's North Slope. These changes include increased runoff through summer and fall and possibly an earlier retreat of the landfast ice (Searcy and others, 1996). In aggregate, both processes will alter the runoff-forced circulation component on the inner shelf in summer and early fall.

The pack ice distribution also is expected to change, in particular, northward ice retreat has increased so that ice-free waters extend over the entire Beaufort Sea shelf and slope from late summer through fall. Diminished ice over the shelfbreak and continental slope should facilitate shelfbreak upwelling (Carmack and Chapman, 2003), which in conjunction with increased light levels due to a diminished ice cover may promote fall phytoplankton blooms.

Persistently Cold Northern Shelf

Previous work has illustrated a north-south transition in the Bering Sea shelf ecosystem (Stabeno and others, 2010). On the southern shelf, biomass is dominated by sub-Arctic and temperate-zone groundfish, with substantial primary production reaching both the pelagic and benthic communities. On the northern shelf, large fish are relatively scarce, and the benthic community receives more of the production than the pelagic community (Stevenson and Lauth, 2012). It has been hypothesized that this north-south transition will be affected by climate warming and that a northward shift in some species will occur as the water column warms. Fish populations on the southern shelf have previously been observed to shift their distributions northward in response to warming (Mueter and Litzow, 2008; Spencer, 2008).

There is now a new understanding, however, of a persistently cold northern shelf, and its associated 'cold pool' (bottom waters ≤ 36°F) that forms beneath the sea ice. Evidence now suggests that the cold northern middle shelf will form a barrier to temperature-limited pollock, cod, and arrowtooth flounder populations, while other, more cold-tolerant populations such as snow crab (*Chionoecetes opilio*) may increase in the north (Stabeno and others, 2012).

Meanwhile, the warming southern shelf is likely to experience profound ecosystem changes. As one window into the cascading effects of warming marine waters, Bering Sea Project scientists have provided one explanation for the surprisingly low recruitment success of pollock during the 2001–2005 'warm' years in the southern Bering Sea. The preferred prey of young-of-the-year (YOY) pollock—large, energy-rich zooplankton—were less abundant during the warm years, and YOY pollock were found to have very low energy reserves before winter (Heintz and Vollenweider, 2010). This likely led to low overwintering survival, resulting in subsequent poor recruitment into the adult pollock population and sharp declines in population size and fishery catches.

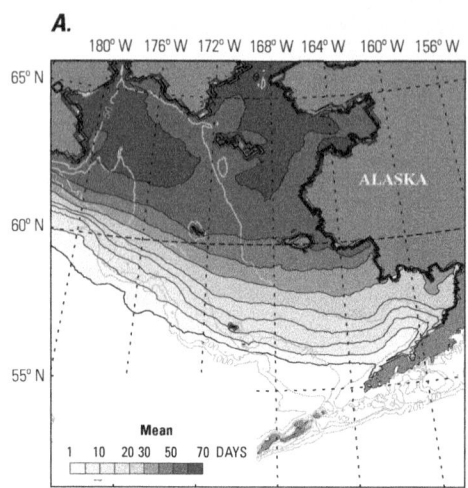

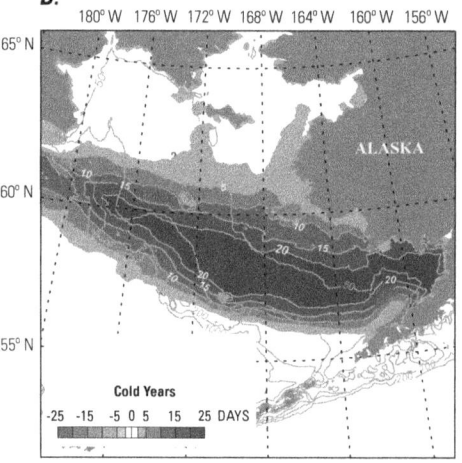

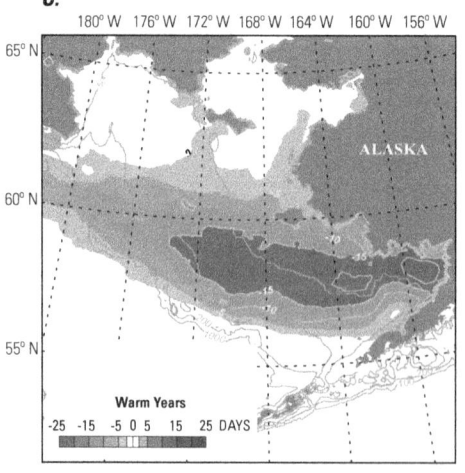

(a)The average number of days in which sea ice was present in March and April during the period 2001–2010. The anomalies of sea-ice coverage during March and April during (b) the cold years, 2007–2010, and (c) the warm years, 2001–2005. [Figure reproduced from Stabeno and others, 2012.]

Sea Ice Conditions

Recent climate variability in the Arctic has resulted in significant sea ice changes in the Bering and Chukchi Seas. These changes are most pronounced in the late summer and early fall, defined as July through October. Meier and others (2007) studied regional variations in Arctic sea ice extent and found that Bering Sea ice extent decreased 39–43 percent in July and October from 1979 to 2006. In the Chukchi region, July–October sea ice extent declined by 24–47 percent. The overall downward trend of Arctic sea ice can be explained from increasing air and ocean temperatures and changing atmospheric and ocean circulation (Stroeve and others, 2011). An anti-cyclonic regime has dominated the Arctic Ocean for more than a decade, resulting in greater movement of older, thicker ice towards the Atlantic Ocean and more rapid melting of sea ice in summer months. In past decades, this anti-cyclonic regime had less effect on Arctic sea ice due to greater ice pack and thickness (Overland and others, 2008), and current sea ice consists of significantly less multi-year ice and summer sea ice extent than in past decades (Stroeve and others, 2011). These two attributes allow the anti-cyclonic regime to transport more sea ice across the Arctic during summer months, which results in faster melt and significantly reduced sea ice extent in the Chukchi Sea (Ogi and others, 2010; Overland and Wang, 2010). These conditions, combined with an extreme Pacific-North American pattern in 2007 led to the extreme minimum sea ice event (L'Heureux and others, 2008), during which little sea ice remained in the Pacific Arctic. In the most recent winters (2008–2011), however, there has been sufficient refreezing to reduce the magnitude of the negative ice extent anomalies during winter in the Bering Sea. The tendency for the extent of sea ice to be reduced more strongly in summer than in winter is consistent with greenhouse-driven climate model simulations.

Current Effects of Climate Change on Polar Bears and Pacific Walrus

Polar bears are dependent on sea ice for much of their life history. Laidre and others (2008) consider them to be one of the most sensitive of Arctic marine mammals to climate change. In the Southern Beaufort Sea, where polar bears commonly give birth to cubs in dens on sea ice, increasing numbers of female bears are now denning on land (Fischbach and others, 2007), presumably in relation to declining conditions on the pack ice. In Western Hudson Bay, sea ice is now absent for 3 weeks longer than just a few decades ago, resulting in reduced body condition of bears and reduced survival of the youngest and oldest cohorts (Stirling and others, 1999). This population is now believed to be in decline (Regehr and others, 2007). Sea ice in northern Alaska now retreats far offshore, resulting in increased numbers of bears coming ashore in the summer and fall (Schliebe and others, 2008). Growth rates of the Southern Beaufort Sea polar bear population are related to the length of time sea ice is found over the continental shelf (Regehr and others, 2009), however, despite massive retreats in sea ice, declines in polar bear populations in Alaska have not been documented. But declines in sea ice, presumably mediated through poorer hunting of ice seals, has resulted in diminished size of polar bears in this population (Rode and others, 2010) as well as in others (Rode and others, 2012).

Compared to polar bears, much less is known about the current effects of climate change on Pacific walrus. Walrus are dependent on sea ice as a platform for birthing, nursing, and resting between foraging trips and they feed on bivalves and other invertebrates on the seafloor (Fay, 1982). Summer sea ice extent in the Chukchi Sea has decreased rapidly in recent years (Douglas, 2010), retreating over deep waters where walrus cannot feed. In response, walrus are coming ashore during the summer in increasing numbers in both Northwest Alaska and Chukokta, Russia. These onshore aggregations are sensitive to disturbance and have resulted in trampling mortality of calves (Fischbach and others, 2009) and may result in competition for food near these coastal haul-outs.

Ocean Acidification

Ocean acidification is an emerging global problem that will intensify with continued CO_2 emissions and may significantly affect marine ecosystems and calcifying organisms.

Since the beginning of the industrial revolution, the average pH of ocean surface waters has decreased by about 0.1 unit—from about 8.2 to 8.1. This change amounts to about 20-percent increase in acidity (that is, concentration of H_3O+ ions), and a similar decrease in the concentrations of carbonate compounds (primarily aragonite and calcite; Mathis and others, 2011) used by calcifying organisms to construct tissues such as skeletons and shells. Models project an additional 0.2–0.3 decrease in pH by the end of the century, even under optimistic scenarios (Caldeira and Wickett, 2005; National Research Council, 2010). This change exceeds any known change in ocean chemistry for at least 800,000 years (Ridgewell and Zeebe, 2005).

The polar ocean is particularly vulnerable to ocean acidification due to relatively low pH and low temperature of polar waters compared to other waters (Orr and others, 2005; Steinacher and others, 2009) and the low buffer capacity of low-salinity waters that result from melting sea-ice (Yamamoto-Kawai and others, 2009). A significant fraction (7.5 percent) of the global net CO_2 uptake (ca. 2200 Tg/yr) has occurred during recent decades in the relatively small Arctic Ocean (about 2.8 percent of the global ocean surface area). The net uptake rate across the Arctic Ocean has been estimated to be between 65 and 175 TgC/yr (Bates and Mathis, 2009). Furthermore, recent changes in the Arctic have intensified the rate of ocean acidification in this region compared to the global ocean. The cold surface waters of the Arctic Ocean absorb CO_2 more rapidly than warmer surface seawater; and increasing temperature (2°F over the past 150 years) of nearly twice the global average has increased melting of Arctic sea ice, which previously inhibited exchange between atmospheric CO_2 and that dissolved in Arctic surface waters. Until recently, the perennial ice cover has prevented significant equilibration with the atmosphere, creating a polar mixed layer that was undersaturated with respect to atmospheric CO_2. Over the last three decades however, melting of more summer sea ice cover has added freshwater to the ocean, has increasingly exposed shelf waters, and has allowed greater CO_2 exchange to occur in these cold waters. The combination of these processes accelerates the rate at which pH and carbonate mineral saturation state decrease. A recent review by Bates and others (2009) has indicated that carbonate mineral saturation state has been observed to decrease with increasing sea-ice melt fraction. Models have projected that the Arctic Ocean will become undersaturated with respect to carbonate minerals in the next decade. However, some recent field results indicate that parts of the Arctic Ocean may already be undersaturated with respect to aragonite in the late summer months, when ice melt is at its largest extent, including a few local areas on the Canadian Archipelago and Beaufort Sea shelves (Bates and

others, 2009; Chierici and Fransson, 2009; Yamamoto-Kawai and others, 2009). Currently (2012), potentially corrosive (relatively low pH) waters are found in the subsurface layer of the central basin (Jutterström and Anderson, 2005; Cheirici and Fransson, 2009; Yamamoto-Kawai and others, 2009), on the Chukchi Sea shelf (Bates and others, 2009) and in outflow waters of the Arctic found on the Canadian Arctic Archipelago shelf (Azetsu-Scott and others, 2010). On the Chukchi Sea, waters corrosive to calcium carbonate ($CaCO_3$) occur seasonally in the bottom waters with unknown impacts to benthic organisms.

Recent observations in the sub-Arctic North Pacific Ocean (Mathis and others, 2011) have already revealed areas of seasonal $CaCO_3$ mineral saturation rate suppression, and new findings show that the eastern Bering Sea likely will be one of the first ocean acidification impact zones for United States national interests. The eastern shelf of the Bering Sea (fig. 31) is a highly dynamic area that is influenced by many terrestrial and marine processes that effect seawater carbonate chemistry with considerable spatial, seasonal and interannual variability in the saturation states of biogenically important $CaCO_3$ minerals (Mathis and others, 2011). The springtime retreat of sea ice, coupled with warming and seasonally high rates of freshwater discharge, creates distinctive horizontal and vertical zones over the shelf, each with their own unique characteristics (Stabeno and others, 1999).

Given the scenarios for pH changes in the Arctic, the Arctic Ocean and adjacent Arctic shelves, including the western Arctic, will be increasingly affected by ocean acidification, with potentially negative implications for shelled benthic organisms as well as those animals that rely on the shelf seafloor ecosystem. While ocean chemistry and the changes caused by increasing atmospheric CO_2 are well understood and can be precisely calculated, the direct biological effects of ocean acidification are less certain and will vary among organisms, with some adapting well and others potentially not at all. Crucial to the Arctic, high-latitude marine organisms often exhibit low metabolic rates and very slow development and growth rates when compared with similar taxa at mid- or low latitudes. Prolonged life histories of these slow growers produce fewer generations that will have opportunities for successful acclimation or adaptation to seawater that will become progressively elevated in dissolved CO_2. Although many physiological processes in diverse organisms may be affected by rising ocean acidity, the declining carbonate mineral saturation states in surface waters, and establishment of corrosive conditions in some regions, may particularly affect high-latitude planktonic and benthic calcifiers. Because of these factors, marine food supplies could be reduced, with significant implications for food production and security for indigenous populations that depend on fish protein. Within the next 100 years, society probably will see significant changes in marine ecosystems and their services as a result of the long-term effects of ocean acidification (Raven and others, 2005).

Observed Rates of Mean Sea Level Change with Respect to the Alaskan Coastline

Observed rates of mean sea-level change with respect to the Alaskan coastline show a range of differences. In the map below, the color scale for the oceans shows mean sea-level rates in millimeters per year determined for 1993–2012 from satellite altimetry [see Salto/Duacs multi-mission altimeter products from AVISO (2012)]. Contours show land uplift and subsidence rates in the same units, based on an update of the dataset of Freymueller and others (2008). Red contours indicate uplift and blue contours subsidence, with a contour interval of 3 mm/yr (0.1 in/yr), approximately the 20-year average global mean sea-level rise rate. Contours are referenced to land surface baseline data points (blue diamonds).

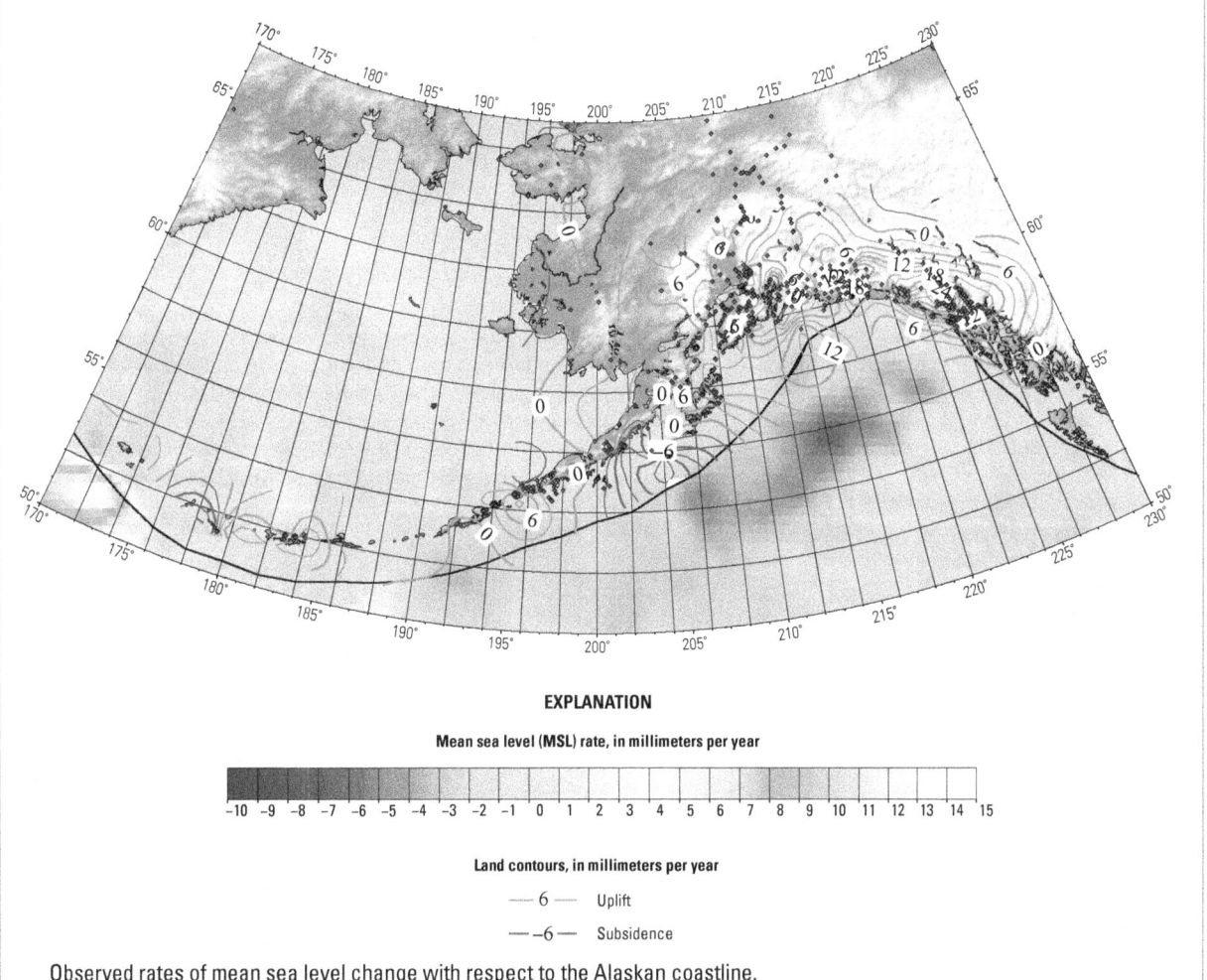

EXPLANATION

Mean sea level (MSL) rate, in millimeters per year

-10 -9 -8 -7 -6 -5 -4 -3 -2 -1 0 1 2 3 4 5 6 7 8 9 10 11 12 13 14 15

Land contours, in millimeters per year

—— 6 —— Uplift

—— –6 —— Subsidence

Observed rates of mean sea level change with respect to the Alaskan coastline.

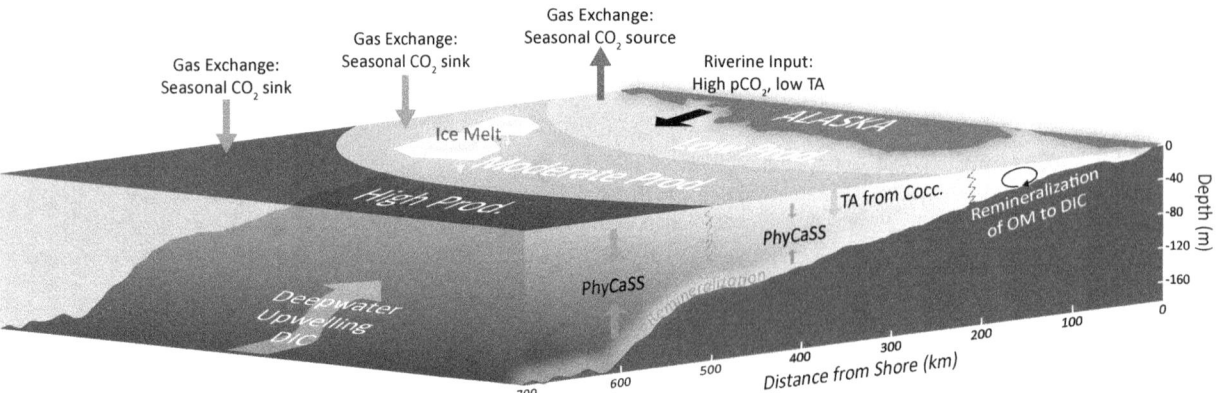

Figure 31.   Generalized description of the processes affecting the carbonate chemistry of the eastern Bering Sea shelf. CO$_2$: Carbon dioxide; pCO$_2$: Partial pressure carbon dioxide; DIC: Dissolved inorganic carbon; Cocc: Coccolithophore; TA: Total alkalinity; PhyCass: "Phytoplankton-Carbonate Saturation State"; OM: Dissolved organic matter.

Sea-Level Change

Relative sea level is the most relevant measure of sea-level change for coastal ecosystems and the built environment, and is defined as the height of the ocean surface (mean sea level) relative to the height of the land surface. Changes in both mean sea level and the height of land surface must be measured or predicted by numerical models in order to assess or predict relative sea level changes. Unlike many parts of the country, changes in both mean sea level and the land surface are important over most of the Alaska coastline.

Regional changes in salinity, water temperature, circulation patterns, and Earth's gravity field produce a large spatial variance in mean sea level rates. All of these effects are significant for sea level rise in Alaska, and they produce a mix of long-term (decades) and short-term (years) effects on mean sea level. Regional mean sea-level change rates for 1992–2009 have been measured by satellite altimetry (see "Observed Rates of Mean Sea Level Change with Respect to the Alaskan Coastline") over most of the Alaska coast, with mean sea-level change either falling or rising more slowly than the global average rate. In the Gulf of Alaska, the ocean surface has been falling at rates of up to a fraction of an inch per year (mostly in the range of 0.04–0.07 in/yr) in coastal regions (Salto/duacs altimeter data from AVISO, 2012). A similar lowering of the ocean surface is observed along the west coast of North and Central America. Ocean circulation changes in response to changes in wind stress at the eastern boundary of the North Pacific have suppressed sea- level rise since the mid-1970s (Bromirski and others, 2011), and interannual variations in ocean surface height are strongly correlated, albeit with a time lag, along these coastal regions (Melsom and others, 2003). Along the Bering Sea coast, mean sea level is rising at rates slightly slower than the global average. There are no such data for the Arctic Ocean, and smaller bodies of water like Cook Inlet are not represented in the altimetry data.

In contrast to the slow rates of mean sea-level change (less than 0.08–0.1 in/yr along the Alaska coast), the land surface moves up or down several times faster along the southern Alaska coast. Presently, emergence or submergence of the coastline is essentially controlled by uplift or subsidence of the land surface. These processes probably are much less important on the Bering Sea and Arctic coasts of Alaska, and oceanic changes are expected to dominate in those regions; however, data of any type for sea-level change in those regions are very limited. The dominant factors controlling uplift or subsidence are glacial isostatic adjustment (Larsen and others, 2005; Elliott and others, 2010) and the tectonic buildup and release of stress related to earthquakes (Freymueller and others, 2008).

In Alaska, the dominant glacial isostatic adjustment effect is from the loss of ice over the last approximately 200 years, after the peak glacial advance of the Little Ice Age (Larsen and others, 2005). Significant subsidence has been observed in a number of river delta systems worldwide, including the Mackenzie in Arctic Canada, but no quantitative data are available for large river deltas in Alaska, such as the Yukon-Kuskokwim delta. Subsidence there is likely occurring, but rates cannot be predicted at this time.

In coastal Alaska, glacial isostatic adjustment causes uplift over a band extending from the northwestern part of the Alaska Peninsula around to southeast Alaska. From the Kenai Peninsula through southeast Alaska, glacial isostatic adjustment causes uplift rates of 0.4 in/yr or more, with the peak uplift rates exceeding 1.2 in/yr (Freymueller and others, 2008; Elliott and others, 2010). These values are about 3 and about 10 times faster than global average mean sea-level rise, respectively. Glacial mass changes in other parts of Alaska are smaller, and uplift or subsidence rates in excess of a fraction of an inch per year are not likely far from presently glaciated areas; however, quantitative testing of the models has not been done for regions outside of southeast Alaska.

Vertical motions also occur because of the buildup of tectonic stress and its release through fault slip in earthquakes. The oceanic Pacific plate is being thrust underneath the North American plate in southern Alaska, causing paired belts of uplift and subsidence. Between earthquakes, there is a band of subsidence that is mostly offshore, and a parallel band of uplift that is mostly onshore (Cohen and Freymueller, 2004). During an earthquake, the pattern is approximately reversed. In Prince William Sound and the Copper River delta, a large area was uplifted in 1964 as a result of slip in the M9.3 Great Alaska earthquake, and the same region has been subsiding at rates of up to 0.3 in/yr since then. In Cook Inlet, the Kenai Peninsula coast subsided in 1964 and has been uplifting since then at rates of 0.3–0.5 in/yr (Cohen and Freymueller, 2004). Rapid tectonic changes in land level persist down much of the length of the Alaska Peninsula, but are more muted in the Aleutians due to changes in the geometry of the plate interface. These motions are thought to be cyclic—over thousands of years the net uplift or subsidence is small, but land level at any given spot is never static.

Future predictions of global sea-level rise are highly uncertain. Recent statistical models predict a mean sea-level rise of 24–63 in. over the 21st century (0.2–0.6 in/yr), several times faster than the present-day rate. Robust estimates of the lower bound on the global average rate are 0.08–0.1 in/yr (8–12 in./century), and there is considerable uncertainty in the upper bound (Jevrejeva and others, 2010). The range of predictions makes it possible that within a few decades mean sea level will be rising at rates that compare to the tectonic uplift rates of 0.3–0.5 in/yr. Even if global mean sea-level rates are closer to the lower bound, it is likely that the Northeast Pacific circulation will change again, when wind stress curl increases as part of a Pacific Decadal Oscillation regime shift; this would result in a period of faster than the global average mean sea-level rise (Bromirski and others, 2011). Thus it is very likely that the ocean surface will be rising faster over the next few decades than today. For much of southeast Alaska, this will simply mean that the land will rise out of the ocean at a slightly lower rate than today, but for most of the Alaska coast the sea-level change trend will become dominated more by submergence rather than emergence. The impact of another major thrust earthquake like the 1964 event may be even more extreme when combined with rise of the ocean surface.

Coastal and Marine Environments

As described previously, projected increases in sea-surface temperatures and reduced sea-ice cover are likely to bring increased storm activity and/or intensity to coastal Alaska (Yin, 2005; Salathé, 2006). Increased erosion is highly likely in coastal areas underlain by permafrost, especially those along the Beaufort and Chukchi Seas (Martin and others, 2009). In addition to the loss of coastline itself, notable effects on coastal habitats likely will include inundation and

salinization of low-lying terrain and alteration of delta habitats (Martin and others, 2009). Barrier islands typically are less than 3 ft above mean sea level (Hopkins and Hartz, 1978), thus making them highly susceptible to erosion. Sea-level rise likely will exacerbate the impact of coastal storms, although the overall impact of sea level will vary with factors such as local elevation, sedimentation rates, and rebound associated with the retreat and thinning of regional glaciers.

Ocean ecosystems and fisheries in Alaska waters are changing rapidly in response to changes in sea temperature and sea-ice conditions (Grebmeier and others, 2006, 2010, 2011). Continued increases in air and water temperatures may lead to a northward shift of fish, bird and marine mammal populations (Grebmeier and others, 2006; Mueter and Litzow, 2008). In more northern waters, changes in sea ice could result in altered plankton production, with cascading effects on fish, bird and marine mammal populations. Loss of sea ice or changes in sea-ice characteristics will directly affect many species (for example, Petersen and Douglas, 2004; Ray and others, 2006).

The shallow northern Bering and southern Chukchi Sea shelf ecosystem off the coast of Alaska is characterized by high, diatom-based primary production in the water column and efficient export from the surface layer to the shallow sediments, feeding a large and diverse benthic community that is critical for benthic-feeding marine mammals and seabirds. Seasonal ice coverage and cold waters have typically limited pelagic fish predation, allowing diving seabirds, bearded seals, walrus, and gray whales to harvest the high benthic production. With recent warming and sea-ice loss, declines in clam populations and diving sea ducks have occurred; large vertebrate predators, such as walrus and gray whales, have migrated farther north; and pelagic fish are expanding their ranges northwards (Moore and Huntington, 2008; Grebmeier and others, 2010; Grebmeier, 2012).

Some of the best-documented examples of biological response to environmental shifts in the Arctic are prey-predator response to hydrographic shifts. For example, a reduction in sea ice has opened up habitat for gray whale feeding farther north from the northern Bering Sea to the latitude of Barrow, Alaska, a likely response to declines in benthic amphipod populations in the historical northern Bering Sea feeding grounds (citations in Grebmeier, 2012). Another change is in dominant clam populations in the northern Bering Sea, which have declined in abundance and biomass, as have Spectacled Eiders that preferentially consume these clams as prey. Modeling by Lovvorn and others (2009) indicates that these diving birds lose more energy resting in the water between feeding bouts than when on ice. Thus both the shift to more open-water conditions and the observed clam population declines are likely key factors creating energy stress for these diving sea ducks. The recent observations of thousands of walrus coming ashore in both the United States and Russia Chukchi coastline are another indication of biological response to rapid sea-ice retreat in the Chukchi

Sea. In addition to the increased mortalities for young walrus on beaches in close proximity to much larger adults, all these shore-based populations have increased energetic requirements to access the productive offshore waters with higher benthic infaunal prey (citations in Grebmeier, 2012).

These changes may influence food web dynamics as well. For example, model projections reveal that phytoplankton primary production will increase in response to greater light availability as a consequence of reduction in sea-ice cover (Arrigo and others, 2008; Arrigo and van Dijken, 2011), although nutrient limitation could ultimately limit the magnitude of this increase (Grebmeier and others, 2010). A shift to smaller algal species sizes has already occurred due to freshening in the western Arctic Ocean (Li and others, 2009), providing another example of potential changes in food web structure and carbon cycling with continued warming.

Distributed Biological Observatory

In order to track biological response to sea-ice retreat and environmental change, a Distributed Biological Observatory (DBO) is being implemented and coordinated by scientists in the Pacific Arctic Group (Grebmeier and others, 2010, 2012; Pacific Arctic Group, 2012). The DBO includes select biological observations at five locations (red boxes) in the Bering and Chukchi Seas for time series measurements to track biological response to environmental change along with coordinated hydrographic measurements. Datasets from 2010 to 2011 pilot studies, along with other biological time series results and updates, are available from the National Oceanic and Atmospheric Administration (2012a). Efforts also are in progress to develop transect-based biodiversity sampling on behalf of the Conservation of Arctic Flora and Fauna– Circumarctic Biodiversity Monitoring Program (2012).

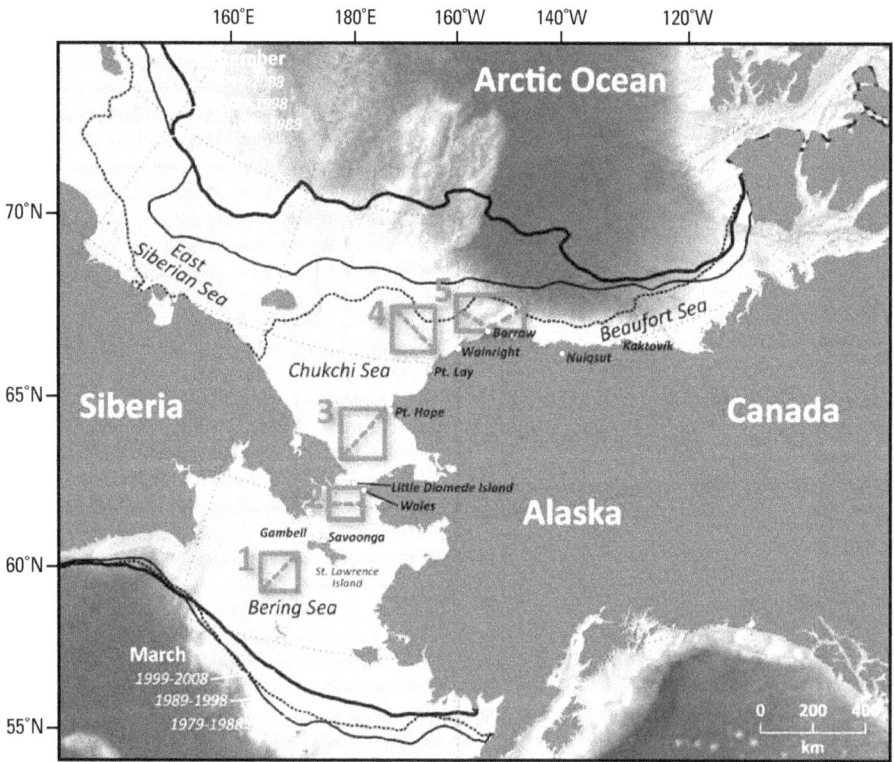

Lines (in black) and dates indicate average sea ice extent for March (maximum) and September (minimum).
Graphic provided by J. Grebmeier (University of Maryland Center for Environmental Science).

Hydrologic Linkages

The hydrologic cycle has a number of direct and indirect linkages between the land and ocean environments. Here we discuss some of the connections and contributions of glaciers and rivers to the ocean environment.

Glaciers

Alaska is home to some of the largest and fastest changing glacier systems on Earth outside the two great ice sheets—Greenland and Antarctica (Larsen and others, 2007b; Berthier and others, 2010). This has direct consequences for local hydrology, sediment transport in glacial streams, freshwater supply to coastal areas, and sea-level change, although these changes are not spatially uniform. Impacts to landscapes and infrastructure may be highly variable, and range from the local to the global scale.

The primary driver of glacier change is climate—specifically temperature (see Oerlemans, 2005). Dynamic glacier response can be stronger than climate forcing for an individual glacier, but the average regional response has been shown to correlate well with climate (Post and others, 2011) and make glaciers one of the prime indicators of ongoing long-term climate change, even though there are usually significant lags in glacier response to climate, ranging from several years to several decades.

Recent estimates of mass loss from Alaskan glaciers range between 40 and 70 Gt/yr (Kaser and others, 2006; Pritchard and others, 2011; Jacob and others, 2012), which constitutes a 0.004–0.007 in/yr contribution to global sea-level rise. The contribution of surplus fresh water to sea-level

rise from Alaska and British Columbia, Canada, glaciers is approximately 8 percent of the ice melt worldwide, or about 20 percent of that contributed by the Greenland Ice Sheet (Jacob and others, 2012). Most of the loss is focused in the Gulf of Alaska (GOA) region, where basin-averaged thinning rates in Alaska's St. Elias Range are a factor of 50 greater than basin-averaged thinning rates in Greenland (Arendt and others, 2009). Although these losses do not force local sea-level rise in the GOA region due to isostatic and tectonic adjustments (see section, "Sea-Level Change"), other coastal communities will be affected (Bamber and Riva, 2010). Arendt and others (2002) compiled the first statewide survey of glacier volume change and estimated a statewide ice loss of 13 mi^3/yr from the 1950s to the mid-1990s, and a doubling of that rate in the following half decade. These numbers were derived by extrapolating small aircraft laser altimetry measurements on up to 100 glaciers in Alaska and British Columbia. More recent estimates have concentrated on the Gulf of Alaska, where the most rapid changes are occurring (VanLooy and others, 2006; Larsen and others, 2007b; Berthier and others, 2010). Two Gravity and Climate Recovery Experiment (GRACE) satellites allow a monitoring of mass changes at a high temporal resolution, but are somewhat limited in spatial resolution (Pritchard and others, 2011). These data document seasonal mass gain and loss as well as the ongoing ice loss from glaciers, and show that the loss rates are highly variable in time (fig. 32).

The influences on the highly productive nearshore marine ecosystems arise from unique physical and chemical properties (for example, turbidity, temperature, nutrient loads) of runoff from glacierized landscapes, which affect both freshwater and marine aquatic habitats (Royer and Grosch, 2006; Hood and

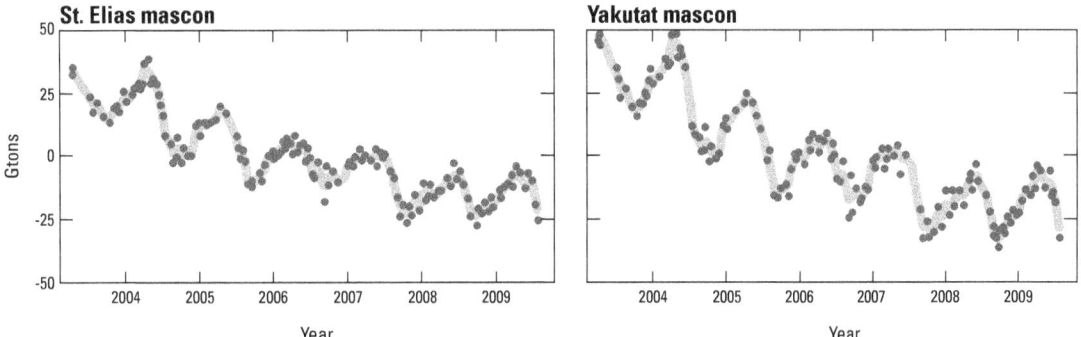

Figure 32.   Mass changes in southeast Alaska glaciers derived from satellite gravimetry. The figure shows the seasonal changes due to the accumulation and melt of snow, and the longer term trends due to the ice loss of glaciers. From Pritchard and others (2011), reprinted from the Journal of Glaciology with permission of the International Glaciological Society. Gtons: Gigatons; mascon: surface mass concentration.

Berner, 2009; Hood and others, 2009; Fellman and others, 2010). As an example, runoff from glaciers along the Gulf of Alaska, which exceeds the annual discharge of the Mississippi River, is estimated to supply approximately 50 percent of the total freshwater discharge into the Gulf of Alaska (Neal and others, 2010). This water generally is delivered at lower temperatures, and with depleted nutrient levels as compared to runoff derived from forested ecosystems. However, discharge from glaciers is increasingly being recognized as an important source of labile organic matter (Hood and others, 2009; Bhatia and others, 2010), phosphorus (Hood and Scott, 2008), and micronutrient iron (Schroth and others, 2011; also see "Glacial-Flour-Derived Dust as an Imporatnt Source of Iron to the Gulf of Alaska") to receiving marine ecosystems. Changes in Alaska's glacier runoff also have the potential to significantly affect the availability of hydropower resources in south-central and southeast Alaska (Cherry and others, 2010), particularly in light of the increased electrical generation potential associated with increasing glacier mass loss.

Glaciers act as reservoirs of water, storing it in the form of snow and ice during cold, wet periods and releasing it during warm, dry periods. Changes in glacial cover profoundly affect seasonal hydrology. During rapid glacier retreat, average river flow is increased, but this trend ultimately reverses as glaciers become smaller. This factor needs to be considered for planning purposes on decadal time scales, such as for hydroelectric installations. A gradual loss of glaciers also leads to reduced water storage as ice and a potential increase in extreme flood events.

Not all glaciers in Alaska change equally. While the vast majority of glaciers in Alaska are losing mass (for example, the Yakutat Icefield; Larsen and others, 2007b), the rates of loss vary greatly, and some glaciers are advancing. The highest variability occurs on glaciers that terminate in water (either sea water or lakes), which imposes a dynamical signature on glaciers that can overwhelm any immediate climate forcing (Post and others, 2011). For example, Hubbard Glacier near Yakutat has been advancing for over a century, and it is anticipated that this advance will continue in the next few decades. Its continued advance is expected to lead to a renewed build-up of a dam that blocks Russell Fjord from the ocean. Such dams have formed twice in historical times (1986 and 2002), and they are almost certain to form again (Motyka and Truffer, 2007). If this closure becomes sufficiently robust to outlast the annual advance/retreat cycle of the glacier front, it will redirect the entire drainage area of Russell Fjord to the Situk River and damage infrastructure and fish habitats in and around Yakutat.

Glaciers terminating in tidewater have the potential to shed ice more rapidly than land-locked glaciers. The single largest volume loss occurs at Columbia Glacier in Prince William Sound, which has been in rapid retreat since the mid-1980s (Post and others, 2011). It serves as an important analogue for many rapidly changing glaciers around the world, including those of outlet glaciers in Greenland (Motyka and others, 2010).

Approximately 55 percent of current rates of global sea-level rise are contributed from the melting of glacier ice (Cazenave and Llovel, 2010). Consensus estimates through 2006 for the mountain glaciers suggest their contribution is 0.06 ± 0.01 in/yr (Kaser and others, 2006). Understanding the climate-induced vulnerability of the freshwater flux from these glacier changes is important, considering that hydrologic changes expected from perturbed glacier runoff are much larger than those projected for other components of the water cycle (Solomon and others, 2007). In Alaska, the large temporal variability of glacier runoff provides physical controls on the structure of the freshwater-driven Alaska Coastal Current (Royer, 1982). Over 5 years, GRACE satellite measurements of the annual change in surface mass within the Gulf of Alaska region vary by more than 60 percent about the mean (Pritchard and others, 2011). As described earlier, the Alaska Coastal Current flows northwestward along the Gulf of Alaska coast until exiting to the Bering Sea at Unimak Pass (Weingartner and others, 2005), demonstrating the connections between Alaska's large coastal glaciers and the Arctic.

River Discharge

Freshwater discharge into the Arctic Ocean is dominated by three large Russian rivers (Yenisey, Lena, and Ob) and the Mackenzie River in Northwest Canada. These four rivers contribute a combined 450 mi^3/yr (Raymond and others, 2007). Freshwater discharge from Alaska into the Arctic Ocean is from the Arctic (79,100 mi^2) and Northwest (68,000 mi^2) regions, which constitute approximately 25 percent of Alaska. On the basis of available streamflow data for about 29 percent of these regions, freshwater runoff to the Arctic Ocean is approximately 9 mi^3/yr. Although this underestimates the total Alaskan freshwater discharge, the true amount represents a small percentage of the total freshwater input to the Arctic Ocean.

Freshwater discharge from Alaska into the Bering Sea is from the Yukon (200,000 mi^2) and Southwest (112,000 mi^2) regions, representing 54 percent of Alaska. The flow of the Yukon River at Pilot Station represents the entire Yukon region, and streamflow data in the Southwest region represents 68 percent of this region. On the basis of available streamflow records, freshwater runoff to the Bering Sea is approximately 76 mi^3/yr.

Within the Yukon region, a positive trend in annual discharge has been noted by Brabets and Walvoord (2009), likely due to increased glacier melting. In addition, the components and timing of runoff have changed. Increases in groundwater levels due to permafrost thawing have been documented as well as earlier spring flows offset by lower summer flows (Walvoord and Striegl, 2007). Altering the portion of groundwater in relationship to the total discharge of water in the river will potentially shift the composition of biogeochemical exports, including dissolved inorganic carbon, dissolved organic carbon, dissolved inorganic nitrogen, and dissolved organic nitrogen.

Freshwater discharge into the Gulf of Alaska comes from the southeastern (40,000 mi^2) and south-central (80,000 mi^2) regions, representing 21 percent of Alaska. Neal and others (2010) determined the freshwater discharge to the Gulf of Alaska from these regions to be 19 mi^3/yr. Freshwater discharge exhibits a strong west to east gradient, with the majority of discharge originating in regions from Prince William Sound to the east. Overall, this high discharge is a result of several factors including: (1) high levels of orographic precipitation, (2) relatively low levels of evapotranspiration, and (3) the predominance of relatively short, steep watersheds within the basin.

Compared to the other regions of Alaska, freshwater discharge in the Gulf of Alaska watershed is heavily influenced by glacier discharge. In total, glaciers cover 18,000 mi^3 or 18 percent of the Gulf of Alaska drainage basin, and account for 50 percent of the freshwater discharge. The changes to be expected in glacier runoff are larger than those generally projected for other components of the water cycle (Solomon and others, 2007). The fact that glacier runoff accounts for 50 percent of total freshwater discharge to the Gulf of Alaska suggests that changes in glacier volume have the potential to substantially alter fluxes of freshwater to the Gulf. Moreover, the fact that annual glacier volume loss along the Gulf of Alaska increased by approximately 9 mi^3/yr during the last several decades of the 20th century (Arendt, 2002) indicates freshwater discharge to the Gulf may be increasing.

The Land Environment

Coastal Erosion Processes

In Arctic settings, both thermal and mechanical processes are important in coastal processes and erosion. Coastal sediments are normally locked in place by ice, and thawing of the ice is necessary before mechanical processes (that is, waves, currents, and wind) can transport the sediments. In addition, shore-fast, bottom fast, and sea ice often protect the coastal zone from wave action. For example, coastal villages on the Chukchi Sea (for example, Shishmarev and Kivalina) were historically protected from the brunt of large late fall storms by nearshore and and off-shore sea ice (U.S. Army Corps of Engineers, 2006). In the last decade, however, these villages have been subject to the full brunt of these storm waves and surge flooding because of diminished sea ice coverage in the late fall.

Two process sequences are responsible for much of the coastal erosion in Arctic Alaska: (1) niche erosion followed by block collapse and (2) thaw slumping. The term "niche erosion" refers to the cutting of a niche at the base of the coastal bluff, and along with block collapse, is a four process sequence (fig. 33).

Storms raise water levels, allowing the sea to directly contact the base of the bluff. Waves and currents thermally and mechanically erode a niche at the base of the bluff. The niche grows until the overburden exceeds the bluff strength

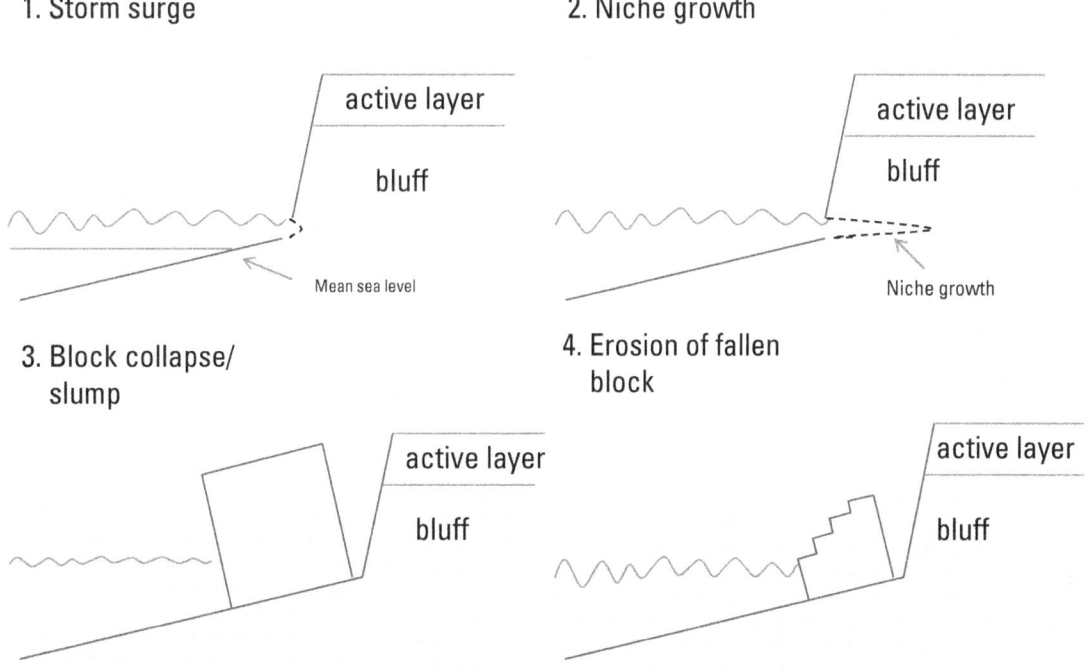

Figure 33.   Conceptual model of the processes responsible for niche-erosion/block-collapse in Arctic Alaska (from Ravens and others, 2012).

Temperature and Coastal Housing Loss

Alaska coasts represent important locations for Native communities, where one of the most immediate impacts arising from temperature change is being felt. These areas are unique among U.S. coastal systems because of the presence of ice, both in the ground as permafrost and on the ocean as sea ice. Increasing temperatures result in permafrost melt, thus destabilizing the ground, and a reduction in the protective ice cover, allowing waves to melt the base of permafrost bluffs and increase coastal erosion. The result is loss of critical infrastructure such as housing.

Photograph by Ned Rozell, Geophysical Institute/University of Alaska Fairbanks.

and block collapse results. The fallen block is then eroded thermally and mechanically by waves and currents. Niche erosion/block collapse is the dominant erosion mechanism in locations dominated by coastal bluffs. Beach survey measurements (Mars and Houseknecht, 2007; Jones and others, 2009b, 2009c) and modeling (Ravens and others, 2012) have documented the rapid and accelerating coastal erosion in locations where this erosion sequence is dominant. Ravens and others (2012) indicate that the main driver in the acceleration of erosion is increasing nearshore water temperature, which weakens the frozen sediments. Recent observations have shown significant erosion even in the absence of storms.

Thaw slumping is predominant in coastal bluffs with appreciable coarse material (for example, on Barter Island, on the northeast coast of Alaska). The term "thaw slumping" refers to the slumping or sloughing of the bluff face following thaw due to radiation or convective heating. Bluff erosion in these areas leaves a significant lag deposit that heightens the beach in front of the bluff and reduces the frequency of niche erosion.

Three main aspects of climate change in Arctic Alaska directly affect coastal erosion: (a) increased temporal and spatial extent of open water and reduced sea-ice concentrations, (b) increased water temperatures, and (c) increased air temperatures. The niche erosion/block collapse sequence will accelerate due to all three of these factors. Increased water temperature, increased wave height due to more open water, and increased temporal extent of open

water [factors (a) and (b)] will intensify and prolong the niche erosion process. These process changes, along with increasing air temperatures [factor (c)], also will accelerate the erosion of fallen blocks. Hence, continued increasing erosion rates due to niche erosion/block collapse are expected. Thaw slumping also will accelerate due to warming of the atmosphere [factor (c)]. However, this process would not be directly changed by oceanic changes.

Permafrost

Permafrost changes affect natural systems, resulting in local, regional and even global feedbacks. In addition, permafrost thaw stretching over years to decades locally affects existing infrastructure and its design, thereby increasing maintenance and construction costs (Larsen and others, 2008). Permafrost thaw also influences freshwater supply for industry and communities (White and others, 2007; Alessa and others, 2008; Jones and others, 2009a).

Two types of permafrost have been identified as most vulnerable to surface thaw in a warming environment in the near future (decadal scale): (1) relatively warm, patchy and thin permafrost in sub-Arctic and boreal regions, much of which is already in imbalance with climatic conditions in interior, western, and southern Alaska and largely protected from thaw by vegetation and soil organic layers (Shur and Jorgenson, 2007); (2) permafrost with high ground ice content (>20 percent excess ice by volume) in near-surface layers and vulnerable to rapid thermokarst and erosion once the ice in these layers starts to melt (Kanevskiy and others, 2011). Only 135,500 mi² (27 percent) of the Alaska permafrost zone is currently (2012) classified as thaw-stable, having low or no ground ice content; the remaining 73 percent (about 370,000 mi²) belongs to permafrost regions with variable-to-high ice content and clear indicators of past vulnerability to thaw, such as presence of thermokarst lakes, thaw slumps, thaw pits, and similar landforms (Jorgenson and others, 2008).

Various disturbances can have direct impacts on permafrost, including wildfires, thermokarst and thermo-erosion (Grosse and others, 2011). Permafrost with high ice content in the upper 33 ft may experience the most dramatic effect from thawing, as melting of irregularly distributed ground ice will result in a decrease in soil volume and thus uneven surface subsidence.

Permafrost change and degradation is far more common within areas of discontinuous permafrost. Near-surface warming in areas of discontinuous permafrost can be as high as 0.4°F/yr, with the result being extensive subsidence as ice-rich ground melts (Hinzman and others, 2005). Resulting thermokarst terrain has been observed throughout the discontinuous permafrost zone, particularly in areas dominated by boreal forest (Jorgenson and others, 2001).

The most direct indicator of stability or changes in permafrost state is the permafrost temperature (Romanovsky and others, 2002). Systematic observations of permafrost temperature in Alaska, Canada, and Russia since the middle of the 20th century allow assessment of changes in permafrost temperatures on a decadal time scale from around the Arctic. Trends in permafrost temperatures and resulting permafrost dynamics across Alaska are broadly consistent with patterns of increasing air temperatures (Hinzman and others, 2005). Observations and model simulations of continuous permafrost surrounding Barrow and at locations throughout the North Slope region show some cooling in the 1950s and 1960s followed by warming that began in the late 1970s (Romanovsky and Osterkamp, 2000). At most permafrost observatories, there was substantial warming during the 1980s and especially in the 1990s in Alaska (Romanovsky and others, 2007; Osterkamp, 2008; Smith and others, 2010; Romanovsky and others, 2010a) and adjacent Northwest Canada (Smith and others, 2010; fig. 34).

North Slope Borehole Temperature Profiles

Because of the harsh operating environment, it is difficult to keep climate-monitoring instruments running in the Arctic without significant data gaps. However, the Earth itself continuously records climate-related energy imbalances at its surface, day-after-day, year-after-year. In northern Alaska, periodic temperature measurements in wells drilled into deep permafrost have captured an on-going warming trend as excess heat is pumped into the ground. On average, near-surface permafrost temperatures have warmed 3–4°C in this region since 1990. With the availability of data from co-located climate-monitoring stations over the last decade, it is known that the recent permafrost warming in northern Alaska primarily is related to warming air temperatures during the snow-covered seasons, especially during winter. Warming air temperatures in this region likely are related to the observed reduction in sea ice in the Bering, Beaufort, and Chukchi Seas in recent years. As ice-rich permafrost warms, it becomes more susceptible to various forms of failure. Coastal erosion rates have doubled along the Beaufort Sea over the last two decades, while slope and riverbank failures have become more common. Degrading permafrost can have significant impacts on human infrastructure (for example, pipelines), ecosystems, and indigenous populations.

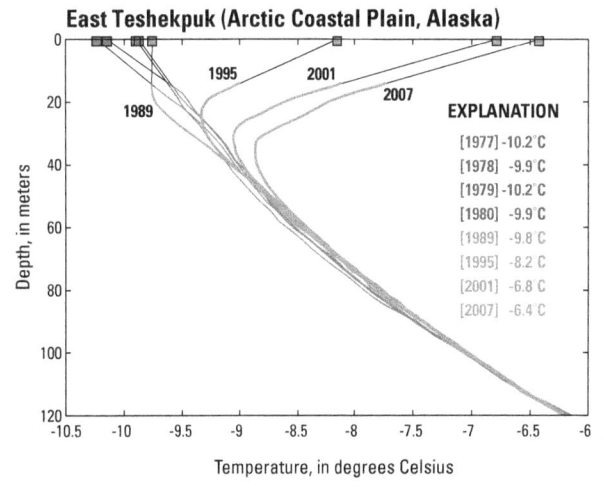

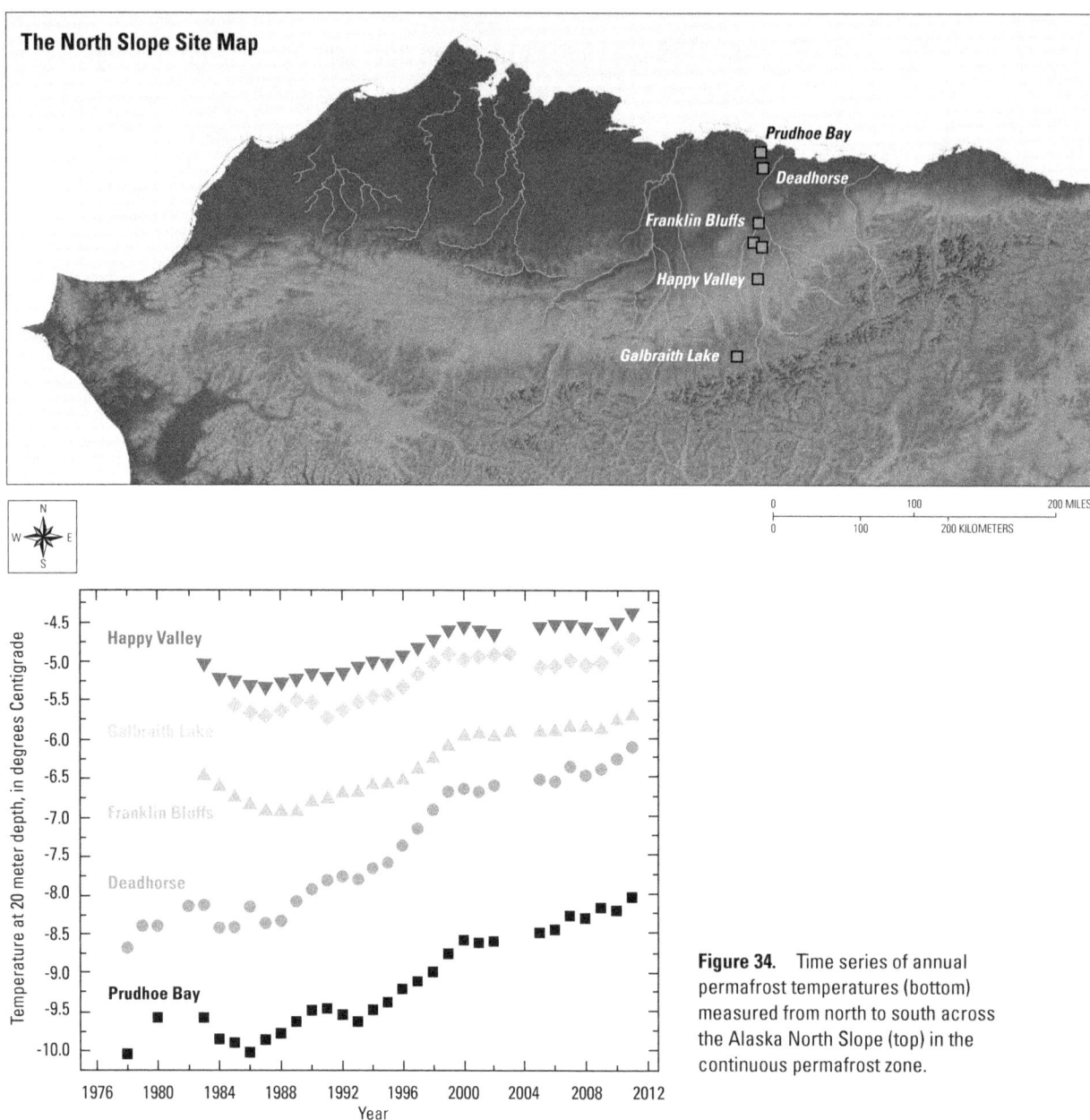

Figure 34. Time series of annual permafrost temperatures (bottom) measured from north to south across the Alaska North Slope (top) in the continuous permafrost zone.

Bore-hole studies from the Arctic coastal plain likewise show a roughly 5°F warming since the late 1980s (Clow and Urban, 2002). At the same time, many areas within the continuous permafrost zone have shown increases in active layer temperatures (Osterkamp, 2003) and a decrease in active layer re-freezing rates (Hinzman and others, 2005).

A common feature at Alaskan sites, similar to other permafrost regions, is more significant warming in relatively cold permafrost[4] than in warm permafrost in the same geographical area (Romanovsky and others, 2010a). The magnitude and nature of the warming at the permafrost table varies depending on region and typically ranges from

---

[4]Cold permafrost: Remains below 30°F / -1°C or as low as 10°F / -12°C. It can take considerable heat without thawing. Warm permafrost: Remains just below 32°F / 0°C. Very little additional heat may cause it to thaw.

0.5 to 7°F over the 1983–2003 period (Osterkamp, 2008). However, during the 2000s, permafrost temperatures have been relatively stable on the North Slope of Alaska (Smith and others, 2010), and there has even been a slight cooling of 0.2° to 0.5°F in interior Alaska from 2007 to 2011 (not shown). During the last decade, continued warming is observed only at near-coastal sites. The latest data may indicate that the observed warming trend along the coast has begun to propagate south towards the northern foothills of the Brooks Range, where a noticeable warming in the upper 60 ft of permafrost has become evident since 2008 (Romanovsky and others, 2011).

In 2011, new record high temperatures at 60-ft depth were measured at all permafrost observatories on the North Slope of Alaska, where measurements began in the late 1970s (fig. 34). The distinct permafrost warming on the North Slope and a slight cooling in interior Alaska is in good agreement with observed air temperature patterns (Overland and others, 2011). These patterns may be additionally influenced by observed changes in snow distribution dynamics (Derksen and Brown, 2011).

The warming of cold permafrost in the continuous permafrost zone generally has not resulted in thawing during the past 60 years, while similar warming in some discontinuous permafrost regions has resulted in widespread thawing (Romanovsky and others, 2010a, 2010b).

Snow Cover

Satellite imagery indicates that the extent of snow cover in the Northern Hemisphere has decreased by about 10 percent since the late 1960s, with stronger trends noted since the late 1980s (Lemke and others, 2007). Similar trends have been documented in Alaska, and National Oceanic and Atmospheric Association (NOAA) snow cover charts (fig. 35) show a statistically significant decrease in the extent of snow cover in May. April also showed a decrease in snow cover extent of 58 mi²/yr between 1970 and 2011, although this trend was not as significant (p = 0.04). The NOAA data did not indicate a significant trend during any of the September–March months. Furthermore, an analysis of existing satellite data, modeled outputs, and re-analysis of snow cover data indicates that since 1972, snow return in the fall has occurred approximately 2 days per decade later and snowmelt has occurred 4 to 6 days per decade earlier, averaged across all of Alaska (Brown and others, 2010).

In-situ observational measurements throughout Alaska generally agree with the remotely sensed data. Stone and others (2002) found that on the basis of observations at several coastal weather stations, the spring snowmelt in northern Alaska near Barrow has advanced by approximately 8 days since the mid-1960s. This change is attributable to decreases in winter snowfall followed by warmer spring conditions. Moreover, the change in snowfall near Barrow is attributable to a greater frequency of northerly airflow during winter that diminishes snowfall over northern Alaska, while an influx of warmer moist air from the North Pacific in the spring increases snowmelt. In the more southerly coastal regions of Alaska, winter temperatures also are influenced by the maritime climate, and are often close to the freezing point of water (32°F), such that snow cover is very sensitive to small changes in temperature, particularly at lower elevations, where the warming influence of the ocean is greatest. In Juneau, average snowfall decreased by 16 in., from 89 to 93 in., between 1943 and 2005 (Kelly and others, 2007).

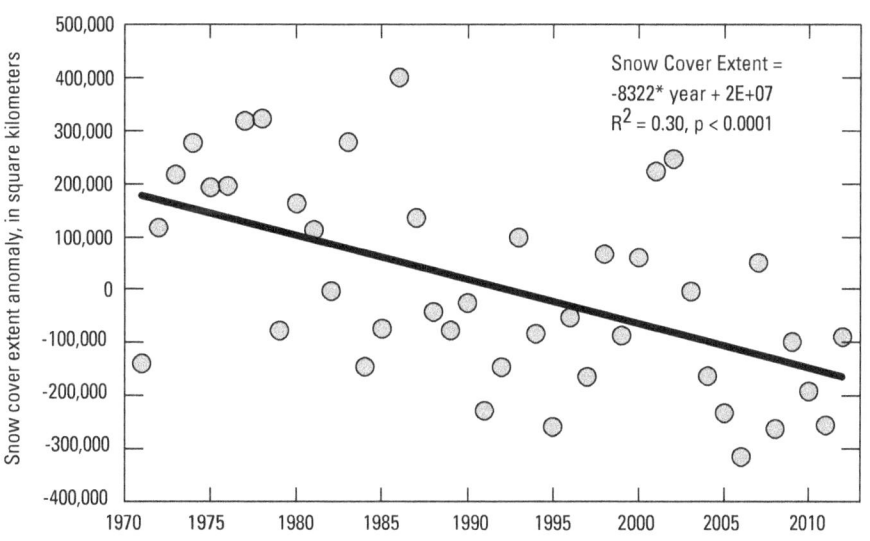

Figure 35. Changes in snow cover extent for May in Alaska based on snow cover charts from the National Oceanic and Atmospheric Association (NOAA). (Data are courtesy of the Rutger's University Global Snow Lab, and are described in Robinson, 1993, and Brown and Mote, 2009.)

During the same period, average winter temperature rose by approximately 2.5°F and average winter precipitation (rain plus snow reported as inches of liquid water) increased by approximately 2.6 in. That is, the decrease in snowfall at sea level appears to be driven by climate warming and a shift in precipitation falling as rain versus snow rather than a decrease in winter precipitation. In Juneau, the change in snowfall regime is most pronounced in April, a month in which snowfall at sea level has become rare. In Fairbanks, from 1906 to 2006, average winter temperatures (for example, October–April) have increased by 1.9°F and precipitation has decreased by 18 percent (Wendler and Shulski, 2009). Although the increase in April air temperature during this period was 5°F, there is no significant trend toward an earlier snowmelt date in April in Fairbanks.

Observed changes in the snow cover duration in Alaska are expected to continue under a warming climate. Based on nine climate scenarios for the years 2003–2100, a modeling study using the Terrestrial Ecosystem Model (Euskirchen and others, 2009) applied to northern Alaska indicates changes in the snow return of 0.13–0.41 days later per year, and changes in snowmelt of 0.08–0.33 days earlier per year (Euskirchen and others, 2009). Similarly, another study, based on a suite of ensembles of 21st century climate projections using the Community Climate System Model Version 4[5], examined changes in snow cover and snow depth in North America and found the largest changes in northern Alaska and northern Canada (Peacock, 2012). This study also found sharp decreases in the number of high snowfall days in Alaska by around the middle of the 21st century. It is important to note that precipitation, and snow in particular, is inherently difficult to model (Frei and Gong, 2005), a problem that is compounded at high latitudes, where there are fewer long-term observational records of precipitation for model-data comparisons.

River and Lake Ice

Few studies have addressed the effect of climate change on river and lake ice in Alaska, but it is vital to understand how projected climate trends will affect these water bodies in the future. In rural, predominantly Native communities, frozen rivers serve as major transportation routes during the winter, allowing an often quick and inexpensive travel between communities and to hunting grounds and food sources. Transportation between villages during periods of open or hazardous ice conditions is confined to expensive air service until appreciable ice and snow cover have formed, allowing the safe use of vehicles (Bieniek and others, 2010). When river and lake ice begin to break-up due to warmer temperatures in the spring, transportation is once again curtailed. In addition, ice jams occurring during spring breakup have produced

nearly all of the record flood crests on the larger rivers. It is apparent that the timing of river freeze-up and the timing and severity of river break-up will be significantly affected by warming climate scenarios (Beltaos and Burrell, 2003).

Determining specific trends in breakup dates is complicated because the existing long-term datasets primarily are based on ground observations at a single point. For example, the trend in the long-term record of breakup dates on the Tanana River at Nenana has been for breakup to occur earlier by several days since the early and mid-20th century (fig. 36). Local meteorological conditions and geomorphology can significantly affect the timing of breakup within a single river, resulting in the potential for contradictory regional trends. Satellite observations of breakup provide for a more regionalized view (Pavelsky and Smith, 2004), but the availability of data for this type of analysis is insufficient to show trends.

In a study of recent trends in Canadian lake ice cover, Duguay and others (2006) found no statistically significant trends in freezeup dates for lakes across Canada. However, trends toward earlier breakup dates dominated, particularly in western Canada, and several of these trends were significant at the 10 percent level. They also found statistically significant trends toward earlier freezeup in eastern Canada and earlier breakups in British Columbia.

Separating the effects of multi-decadal climate teleconnections from long-term climate change is another challenge when examining trends in breakup dates. Although the North Atlantic Oscillation (see National Oceanic and Atmospheric Administration, 2012d) had a significant effect on the January–March temperatures in the Baltic area of Northern Europe, climate forcings related to elevated CO_2 levels and regional spring warming trends were a more significant factor leading to earlier breakups over the last several decades (Yoo and D'Odorico, 2002).

The severity of breakup induced ice jam flooding results from a combination of factors, including ice thickness, snowpack, and spring temperatures. Bieniek and others (2010) found that breakup on four major Alaskan rivers tends to occur earlier when spring (April–May) surface air temperatures are above normal, with increased winter precipitation having a secondary impact by increasing spring river discharge. In neighboring Canada, however, Pavelsky and Smith (2004) found that breakup was associated more strongly with sea-surface temperature anomalies that were tied to the Pacific Decadal Oscillation Index (see National Oceanic and Atmospheric Administration, 2012e). The increase in spring temperatures could result in more thermal breakups that generally are not linked with severe ice jam floods, but it also could lead to scenarios in which a cool spring is followed by a sudden warm-up when a large portion of the snowpack melts and enters the river in a short period. This scenario

---

[5]The Community Climate System Model is a coupled Global Climate Model maintained by the National Center for Atmospheric Research developed by the University Corporation for Atmospheric Research. The coupled components include an atmospheric model, a land-surface model, an ocean model, and a sea ice model.

Breakup Dates: Tanana River at Nenana

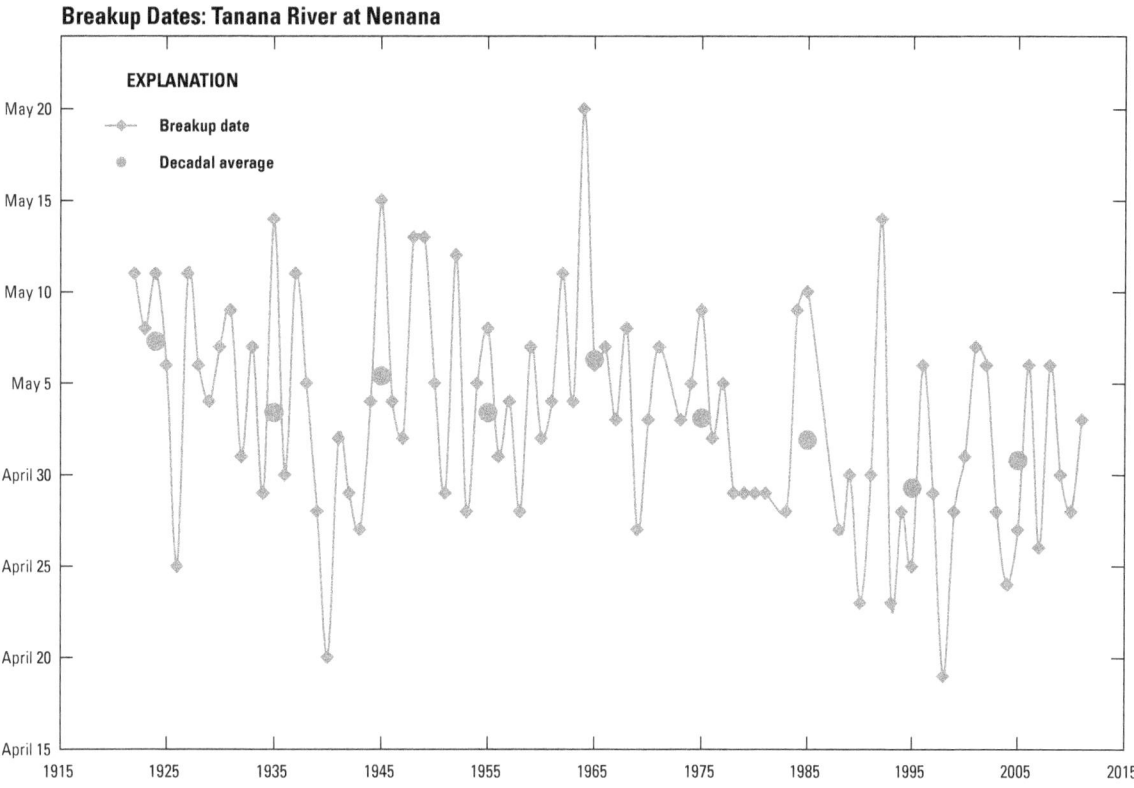

Figure 36.   Breakup trends on Tanana River at Nenana, Alaska (Nenana Ice Classic, 2012).

often results in a dynamic breakup associated with historically severe ice jam flooding such as the Yukon River experienced in 2009 (Janowicz, 2009). Severe flooding also occurred in 2011 on the Kuskokwim River at Crooked Creek, which may indicate that current climate change trends for warmer temperatures and earlier breakups may not translate into a lower risk of flooding from ice jams.

Earlier breakup and freezeup, and thinner river and lake ice have important societal implications that include the following:

- Major transportation routes, such as the ice road on the Kuskokwim River, are available for a shorter period of time and dangerous holes and thin ice will persist much longer into the winter season.

- Fall breakups occur more frequently, such as that on the Kuskokwim River in November 2010, leading to a significant delay in ice road formation, and potentially affecting the subsequent spring breakup (Crooked Creek flooding in May 2011; University of Alaska, Fairbanks Alaska Center for Climate Assessment and Policy, 2011).

- Changes in the thickness and duration of lake and river ice cover may have consequences for both the natural environment and human activities (Prowse and Beltaos, 2002; Lemke and others, 2007; Bieniek and others, 2010; Herman-Mercer and others, 2011), including:

  - Extreme flood events caused by ice jams

  - Interference with transportation and energy production

  - Low winter flows and associated ecological and water quality consequences

  - Potential effects on migratory birds and salmon

  - Lake and river drowning due to unsafe travel over ice

  - Community flooding due to ice jams

  - Changes in river channels

Water Resources

Terrestrial regions of the Arctic would be classified as having high surface wetness during the summer months. This primarily is because permafrost inhibits vertical drainage, and a snowpack that has accumulated over a 7–9 month period ablates in a relatively short period of time. The potential hydrologic pathways of this surface water are evapotranspiration and runoff. In Alaska, over a 5-year period, evapotranspiration constituted 35–50 percent of water exported out of three Arctic watersheds while the remainder of the water exited these watersheds as surface runoff (Kane and others, 2000). During the summer months, latent heat fluxes dominate in wet sites and sensible heat fluxes dominate in dry sites (Mendez and others, 1998). Because of the latent heat required to melt ice in the active layer and the ongoing surface evapotranspiration, the active layer depth is minimized and soil temperatures remain cool. Currently (2012), there are large annual fluctuations in surface wetness; this depends on summer precipitation and climate, with warm summers producing more evapotranspiration than cooler summers.

Hydrologic changes witnessed in the Arctic include drying of thermokarst ponds in Alaska (Yoshikawa and Hinzman, 2003; Jorgenson and others, 2006; Riordan and others, 2006; Roach and others, 2011), in Canada (Smol and Douglas 2007), and in Siberia (Smith and others, 2005), increasing the importance of groundwater in the local water balance, and differences in the surface energy balance. By far, the most significant changes occur in response to changing extent or thickness of permafrost. As permafrost becomes thinner, the sub-permafrost groundwater becomes more important, either by contributing the groundwater to streamflow, or by allowing surface water to drain. The important implications are that in regions underlain by thin permafrost (approximately <60 ft), surface ponds may shrink and surface soils may become drier as the permafrost degrades. In areas underlain by thick permafrost, massive icewedges could degrade, resulting in catastrophic draining of lakes and wetlands.

Precipitation (rain and snow) is a climate parameter that is difficult to measure and complex to predict in the Arctic. The Arctic Climate Impact Assessment (2005) suggests that a 1-percent increase in precipitation per decade was probable over the last century. Most of the climate stations reported by Hinzman and others (2005) showed an increase in annual precipitation over the length of their record (since late 19th century or more recent), yet the summer surface-water balance (precipitation minus potential evapotranspiration) measured in Alaskan North Slope villages decreased since 1960 (Oechel and others, 2000). Seasonal distribution of precipitation is important to consider as winter precipitation is projected to increase with continuing climate change (Serreze and others, 2000), although recent data and model syntheses demonstrate a general decrease in winter precipitation (Liston and Hiemstra, 2011). Although regional increases in winter precipitation were reported, the predominant changes in snowpack for the past 30 years were decreases in snow water equivalent (Liston and Hiemstra, 2011).

The growing season in Alaska appears to be lengthening as snow melts earlier in the spring and arrives later in the fall. Longer summers in Alaska have the potential to be beneficial for the growth of plants. However, the satellite record suggests that the response of plant growth to warming differs in different regions of the State, with growth increasing in the tundra of northern Alaska and decreasing in the boreal forest of interior Alaska (Jia and others, 2003; Goetz and others, 2005; Verbyla, 2008). Analysis of forest growth data indicates that growth in white spruce forests of interior Alaska is declining because of drought stress (Barber and others, 2000; McGuire and others, 2010), and the potential exists for continued warming that could lead to forest dieback in interior Alaska (Juday and others, 2005). The drying of interior Alaska also suggests that agriculture in Alaska may not benefit from longer snow-free growing seasons.

Researchers also have documented a net decrease in the area of both open and closed-basin lakes (that is, lakes with and without stream inputs and outputs) during the latter half of the 20th century in portions of the southern two-thirds of Alaska (Klein and others, 2005; Riordan and others, 2006; Roach, 2011; Rover and others, 2012), but the direction, rates, and magnitudes of lake area change are heterogeneous throughout the State (Roach, 2011). In interior Alaska, the decrease in lake area appears to be caused by the conversion of lake shores to peatlands (fig. 37) because of accelerated permafrost thawing and lengthening of the growing season. The increase in growing season facilitates increases in plant growth, floating mat encroachment (fig. 38), transpiration rates, and the accumulation of organic matter in lake basins. Concurrently, lake area increase is often associated with lateral permafrost degradation (Roach and others, 2011). In south-central Alaska, however, a large number of water bodies have shrunk in response to warming since the 1950s, and subsequently have been invaded by woody vegetation (Klein and others, 2005; Berg and others, 2009). The combined effects of wetland drying and vegetation succession in this region have resulted in wetlands becoming weak carbon sources rather than strong carbon sinks, which has important consequences for the global climate system.

The loss of area in closed- and open-basin lakes also may be indicative of a lowering of the water table that has the potential to convert wetland ecosystems in interior Alaska into upland vegetation. A substantial loss of wetlands in Alaska may have profound consequences for management of natural resources on National Wildlife Refuges in the State, which cover more than 77 million acres and comprise 81 percent of the National Wildlife Refuge System. These refuges provide breeding habitat for millions of waterfowl and shorebirds that winter in more southerly regions of North America, and may present a substantial challenge for waterfowl management across the National Wildlife Refuge System (Griffith and McGuire, 2008). Wetland areas also have been traditionally important in the subsistence lifestyles of Native peoples in interior Alaska, as many villages are located adjacent to wetland complexes that support an abundance of wildlife subsistence resources. Thus, the loss of wetland area has the potential to affect the sustainability of subsistence lifestyles of indigenous peoples in interior Alaska.

Figure 37.   An example of progressive lake drying in boreal forest wetlands within Yukon Flats National Wildlife Refuge, Alaska. (Photograph by May-Le Ng).

Figure 38.   Boreal forest wetlands within Yukon Flats National Wildlife Refuge, Alaska, with examples of floating mat encroachment in the foreground. (Photograph provided by Jennifer Roach from http://www. marysrosaries.com/collaboration/index.php?title=Category:Images).

Wildfire

The area burned in the North American boreal region has tripled from the 1960s to the 1990s due to increased frequency of large fire years (Kasischke and Turetsky, 2006). Since 2000, interior Alaska has experienced four large fire years (years in which more than 1 percent of the landscape burned) in which 17 percent of the landscape burned (Kasischke and others, 2010). It is estimated that these fires reduced the coverage of coniferous black spruce forest by 4.2 percent and increased the coverage of broadleaf deciduous forest by 20 percent (Barrett and others, 2011). By the end of this century, area burned is projected to triple in Alaska for a climate scenario of moderate rates of increase in fossil fuel burning (based on the B1 emissions scenario), and to quadruple for a climate scenario of high rates of increase in fossil fuel burning (based on the A2 emissions scenario; Balshi and others, 2009; Trainor and others, 2009). Increases in area burned would increase fire risk to rural indigenous communities and reduce subsistence opportunities, and has implications for fire suppression/fighting policy (Chapin and others, 2008). Collaborations between communities and agencies to harvest flammable fuels for electrical power generation near communities and to use wildland fire for habitat enhancement in surrounding forests could reduce community vulnerability to both direct and indirect effects increased wildfire.

Alaska's fire regime and land cover are projected to transform further as rising temperatures amplify insect outbreaks (Arctic Climate Impact Assessment, 2004; Wolken and others, 2011). Analyses of historical insect and fire disturbance in Alaska indicate that the extent and severity of these disturbances are intimately associated with climate (Duffy and others, 2005; Juday and others, 2005; Balshi and others, 2009). Areas that experience the death of trees over large areas of forest are vulnerable to wildfire as the dead trees are highly flammable. This is particularly a concern in Alaska where fire extent has been increasing in recent decades (Balshi and others, 2009). Alaska communities are experiencing the socio-economic reality of rapid changes in the fire regime and land cover. While costly suppression efforts can stress limited financial resources, these expenditures also provide an important boost to the cash economies of rural communities (Calef and others, 2008; Trainor and others, 2009). As Alaska's human population continues to grow and expand into rural forests, so will human-wildfire interactions (that is, more human ignitions, more area requiring suppression). Wildfire may help subsistence by rejuvenating the forest (Yarie and Van Cleve, 2006) and enhancing habitat for some key subsistence resources such as berries, mushrooms (Nelson and others, 2008), and moose (Maier and others, 2005). However, wildfire may reduce habitat for other primary resources (for example, caribou; Rupp and others, 2006; Joly and others, 2010) and reduce access to harvest areas. Fallen trees following a wildfire block trails used for subsistence, and thick regrowth can hinder travel and obstruct vision of subsistence hunters (Brinkman and others, 2011a). Further, wildfires near rivers and streams temporarily release more ash and woody debris into the water, which can disturb fish habitat (Nelson and others, 2008), tangle or destroy fishing nets, and alter the navigability of rivers used for subsistence (Brinkman and others, 2011b).

Under a higher emissions scenario (A2), which is appearing more likely (Nakicenovic and others, 2000; Raupach and others, 2007), area burned by wildfire may quadruple by 2100. However, when scenarios are modified to account for new successional pathways and altered return intervals, wildfire activity may stabilize toward the end of the century. For instance, model simulations indicate that higher severity fires could lead to a boreal region dominated by deciduous forest, reducing the overall flammability of the forest (Barrett and others, 2011; Johnstone and others, 2011).

Alaska's wildfire regime and land cover change plays a significant role in the global climate system (Wolken and others, 2011). Burning forests contribute directly to global carbon emissions (French and others, 2004; Tan and others, 2007; Zhuang and others, 2007; Balshi and others, 2009). In addition, wildfire may accelerate the rate of permafrost thaw, and recent expert surveys suggest that permafrost thaw will release approximately the same amount of carbon as deforestation (Schuur and others, 2011). Moreover, the effect on climate may be 2.5 times greater than deforestation because permafrost emissions include methane, which has a greater warming effect than does CO_2. The large 2007 tundra fire on Alaska's North Slope released approximately as much carbon into the atmosphere as the entire Arctic tundra biome has stored in the previous 25 years (Mack and others, 2011). Although overall warming, drying, and associated positive feedbacks are likely to outpace negative feedbacks, these interactions should be accounted for. For example, a boreal region dominated by early-succession deciduous forest can uptake more carbon and has greater albedo than does a conifer forest (Euskirchen and others, 2010).

The Human Environment

Alaskan Native Observations of Climate Change Impacts

Alaska Natives are experiencing cumulative effects of climate change. Increased erosion and flooding put Tribes at risk in their traditional homelands; changing ecosystems affect food security, the health of traditional plant and animal species used for food, and traditional ways of life; changing snow, ocean, river, and lake ice conditions make travel more difficult and dangerous; and drier, hotter summers contribute to health problems due to smoke from increased occurrence and size of wildfires.

Through public listening sessions and interviews with Alaska Natives, a number of comments were recorded that speak to both changing ecosystems and the use of traditional knowledge in documenting past and current effects of a changing climate across the Alaskan landscape; summaries of these sessions and interviews are found in appendixes C and D with some highlights noted here. Additional information on the impacts of climate change on Alaska Natives can be found in the other technical inputs to the National Climate Assessment.

At the Alaska Forum on the Environment (2012), a number of speakers described recent past (30–50 years ago) and current observations of changes on the landscape (appendix C). These included advancing spring plant phenology in the Slana region; changes in how fish [salmon] migrate up streams (New Stuyahok area); presence of orca in the Nushagak River; sea-level change; increased growth of willow; appearance of new species of birds in remote areas, and new species of fish (Arctic whitefish) in salmon streams (Barrow); increase in abundance of bees and northern extension butterflies (Kalskag area); the disappearance of fish species, such as burbot (New Stuyahok area, Bristol Bay region); and increased interactions between humans and wildlife (bears, moose, wolves, and coyotes in middle and western Yukon/Kuskokwim area).

Discussions about the use of local Native observations in partnership with western scientific methods take place frequently. Observations that parallel processes in the scientific literatures are given in appendixes C and D. For example, the increasing occurrence of shrubs and trees on hillslopes as reported by Margie Hastings in the Bristol Bay region is similar to that reported by Sturm and others (2001) and Tape and others (2006). Earlier spring thaw and later fall freeze dates reported by Frank Pokiak of the Invialuit Game Council in northwest Canada support work reported by Goetz and others (2005). In southeast Alaska, changes in glaciers and the revegetation of recently de-glaciated areas, as reported by Jo-Ouaack John Morris, from southeast Alaska is similar to those reported by Reiners and others (1971) and Larsen and others (2007b).

In southeast Alaska, modern Tlingit have observed recent changes that include a decrease in the amount of glacial ice in the ice fields and waterways, an increase in air and water temperatures, a decrease in snowfall accumulation, terrain uplift from isostatic rebound, variable effects on fluvial discharge, and changing availability and quality of wood for traditional carving. This region has experienced climatic cycles that have helped to define the elements of life in that area. The observation of these cycles has been handed down within families for generations and it is this traditional knowledge that also can be used to look at past changes across the landscape and its effect on local inhabitants. For example, a study by Connor and others (2009) used Huna Tlingit stories of the Glacier Bay region, combined with radiocarbon dating from geologic evidence, to recreate the historical landscape and model the extent of the ice from past centuries and its effect on the local Huna Tlingit people.

Alaska Natives frequently provide the first indication of climate change impacts and are not just anecdotal verifiers of the findings of western researchers. Local observers apply traditional ecological knowledge to identify whether an occurrence is unusual or significant. In this way, local observers provide invaluable surveillance for change. Through improved communication with researchers (via the Local Environmental Observer program; see Alaska Native Tribal Health Consortium, 2012c), Alaska Natives can assist in describing events and impacts, help to identify research priorities, and participate in research activities and monitoring. In addition, by exploring Alaskan Native descriptions of past events (through stories of past history) and observations of the changing environment, confirmation of both past and current events may be made. It is important to directly engage Native people and perspectives in scientific investigation, vulnerability assessment, and climate adaptation planning. This will ensure that the observations of climate change by Alaska Natives and the impacts of these changes on food sources and traditional way of life are accurately included in analysis.

Human Health

Climate warming in the Arctic currently (2012) has broad implications for human health, affecting vulnerability for injury and disease, water and food security, mental health problems, and damage and disruption to water and sanitation infrastructure (Berner and Furgal, 2005). Little has been published about climate-health connections in Alaska, but the environmental connections and mechanisms are better described. Increasingly Alaskans are experiencing unusual weather events, disrupted landscapes, new travel hazards (National Assessment Synthesis Team, 2000; U.S. Arctic Research Commission Permafrost Task Force, 2003), and interruptions to food harvest and preservation practices (Loring and Gerlach, 2010; McNeeley and Shulski, 2011;

Moerlein and Carothers, 2012). Some positive health effects also are documented, including a longer season for growing healthy foods (Weller, 2005). In light of the growing climate change impacts in Alaska, health organizations are assessing effects, describing the risks and benefits, identifying affected populations, and assisting in the development of appropriate adaptation strategies (Brubaker and others, 2011).

Problems intrinsic to many rural Alaska communities, such as the lack or failure of adequate drinking water systems, sanitary sewage disposal, and usable land, have been negatively affected by climate change (Warren and others, 2005; State of Alaska, 2008). A review of statewide flood and erosion data identified 25 communities likely to face near term climate related impacts to their water and wastewater infrastructure (Alaska Department of Environmental Conservation, 2010a). Other climate effects include warming Arctic lakes in which algae blooms are diminishing water quality and increasing the cost of treatment (Brubaker and others, 2010). Permafrost thaw is causing some lakes to drain completely (Karl and others, 2009b), raising concerns about water availability. Subsidence and erosion are causing widespread infrastructure damage, in some cases interrupting services for months at a time (Brubaker and others, 2011). The implications go beyond increased operations and maintenance costs, and include the potential for increased rates of injury and disease. The relation between piped water service and rates of skin and respiratory infections in rural Alaska communities is well described (Gessner, 2008; Hennessy and others, 2008).

Erosion is causing some shorelines to retreat by tens of feet per year (Karl and others, 2009b). Storm surge in coastal areas, exacerbated by delays in fall sea ice development, has severely damaged facilities in Newtok, Alaska, requiring relocation of the entire community (State of Alaska, 2008). A challenge in developing new facilities and communities is how to design structures for a rapidly changing environment. New guidelines are needed to develop health infrastructure that is resilient, sustainable, affordable, and that meets community needs now and in the future. In the meantime, communities are increasingly vulnerable to failures and related waterborne, vector-borne, and sanitation-related diseases, as well as exposure to environmental contaminants (Macdonald and others, 2005; Alaska Department of Environmental Conservation, 2010b; Loring and others, 2010; Schuster and others, 2011).

Climate change is being linked to changes in forage and vegetation, expansion of the geographic range of animal species, and insect vectors that raise the risk of emerging diseases and invasive species to the northern climate. Examples of climate-related diseases that can be spread from wildlife to humans include leptospirosis, toxoplasmosis, and tularemia (B. Gerlach, Alaska Department of Environmental Conservation, written commun., 2008). Examples of climate-related human diseases include botulism, echinococcosis, giardiasis, paralytic shellfish poisoning, and gastroenteritis.

A further example of climate-related human disease is *Vibrio parahaemolyticus*, bacteria that causes gastroenteritis and typically is associated with the consumption of raw oysters gathered from warm-water estuaries. In the summer of 2004, the first documented outbreak in Alaska occurred and was associated with the consumption of raw oysters (McLaughlin and others, 2005). All of these oysters were harvested when mean daily water temperatures exceeded 59°F (the temperature above which *V. parahaemolyticus* bacteria can proliferate in oysters). Between 1997 and 2004, mean water temperatures in July and August at the implicated oyster farm increased 0.38°F per year; 2004 was the only year during which mean daily temperatures did not decrease below 59°F. The outbreak extended by 600 mi the northern most documented source of oysters that caused illness due to *V. parahaemolyticus*.

Climate change is having significant effects on the availability of key marine and terrestrial species used as food sources, by shifting the range and abundance of species, such as salmon, herring, char, cod, walrus, seals, whales, caribou, moose, and various species of seabirds (Weller, 2005). Decreased availability can negatively affect health, especially when it results in dietary change. Changes in harvest practices also can occur when concerns are raised over food safety or when conditions interrupt traditional methods for food preparation or preservation. In some communities, residents are seeking new methods for food storage, as warming temperatures and thawing permafrost cause failure of traditional methods of food preservation (Brubaker and others, 2009a; Moerlein and Carothers, 2012). Shifting from a traditional to a Western diet is associated with increases in "modern diseases," such as obesity, diabetes, cardiovascular disease, and cancer (Parkinson, 2010). Alaska Natives depend on subsistence for food and for sustaining cultural traditions. When unable to harvest sufficiently, they become vulnerable to negative social, cultural, economic, and nutritional effects (Weller, 2005).

New, more robust and climate-sensitive systems for health assessment and surveillance are needed to address emerging threats and to monitor adaptation strategies. A recent retrospective review of three independent patient databases in Alaska underlines potential climate-to-health connections and the importance of improved surveillance systems. In 2006, Fairbanks medical facilities experienced a four-fold increase in patient visits for wasp stings (compared to 1992–2005); and the first deaths in Alaska associated with insect sting-related allergic reactions. A review of the Alaska Medicaid database from 1999 to 2006 showed statistically significant increases in medical claims for insect reactions in five of six regions, with the largest percentage increases occurring in the most northern areas (Demain and others, 2009). Cold temperatures are a limiting factor for survival of stinging insects, and recent warm winters may be increasing survival of wasp populations, and consequently sting-related patient visits.

Climate change is often not the sole cause of increases in climate-sensitive health outcomes, but interacts with other public health stresses (Gessner, 2008). In Alaska, the protection of human health requires better understanding of health effects, identification of vulnerable populations (Loring and Gerlach, 2009), improved systems for human and wildlife health surveillance, adaptation approaches that are community specific, and infrastructure systems that are appropriate, resilient, and sustainable.

Subsistence

Rural livelihoods in Alaska are tightly connected to climate, weather, and ecosystems. Northern people have relied for millennia on the landscape for their food through a variety of subsistence activities including hunting, herding, gathering, fishing, and small-scale gardening. The importance of wild fish, whether anadromous species such as salmon or non-anadromous species such as whitefish and pike, is the notable constant from south to north in rural Alaska. Use of terrestrial small and large game (including moose, caribou and black-tail deer), waterfowl, and marine mammals differs from south to north and from east to west across the State. For a reader interested in more background on subsistence harvest in Alaska, see Nelson (1969, 1986); Norris (2002); and Wolfe (2004).

The impacts of unexpected changes and unprecedented environmental conditions on the harvests of these subsistence foods are easily observed, and residents of rural Alaska are already reporting unprecedented changes to the distribution and abundance of fish and game. When combined with social and economic change, climate, weather, and changes in the biophysical system interact in a complex web of feedbacks and interactions to make life in rural Alaska more challenging.

Regional climatic and environmental changes are already having a notable, although unpredictable and often non-linear effect on subsistence activities, through changes in hydrology, seasonality and phenology, land cover, and fish and wildlife abundance and distributions (White and others, 2007; Loring and Gerlach, 2009; McNeeley, 2009; Rattenbury and others, 2009; Loring and others, 2011). Despite the broadly scaled directional trends observed and projected for warming and drying in the region (Chapin and others, 2006), the effects of climatic change are being experienced not directionally but in terms of greater inter-annual and inter-seasonal variability (Bryant, 2009; Rattenbury and others, 2009; Wendler and Shulski, 2009). Uncertainty is high regarding how seasonal conditions will play out in the future (Lawler and others, 2010). The timing of the seasons, for instance, including fall freeze-up and spring break-up, are shifting in unpredictable ways from year to year (Mills and others, 2008; Mundy and Evenson, 2011); river ice conditions also are changing; winter ice is thinner and less predictable, and variability in precipitation and snow pack will affect water levels in both the fall and spring (Euskirchen and others, 2007; Hunt and others, 2008; Wendler and Shulski, 2009). Events like an early break-up, or infrequent meteorological events, such as rain-on-snow, can be devastating for caribou and reindeer populations and thus have tremendous impacts on herders and hunters (Rattenbury and others, 2009).

Climatic and environmental changes also can directly influence hunting activities, including transportation across the landscape, and concerns about the spoilage and storage of meat (Brubaker and others, 2009b; Loring and others, 2011). High water levels, fire, and permafrost thaw slumps are all examples of recent changes that raise safety concerns and limit access to traditional harvest areas (Loring and Gerlach 2009; Kofinas and others, 2010). In addition, as ecosystems and seasonal patterns change, the environmental cues that hunters use to predict the weather and location of animals may become less reliable (McNeeley and Huntington, 2007) or out of sync with existing hunting regulations (McNeeley, 2012).

Salmon, which could be described as the cultural keystone food of Alaska, has likewise become a less dependable subsistence resource than in the past, and this has direct implications for food security, especially as one moves up the Yukon River to the Canadian border (Loring and Gerlach, 2010). A closure of the king salmon fishery on this river in 2009, for example, resulted in empty storage facilities, empty smokehouses, and barren fish racks from Stevens Village up through Fort Yukon. The 2009 closure produced a food security crisis, especially in combination with low harvest rates of moose and other terrestrial resources in some areas, the high price of fuel, and climate-driven changes in hydrology and water resources (Loring and Gerlach, 2010).

Climate-related changes in sea ice and weather patterns also are already creating numerous new environmental challenges for those who harvest marine species. Surface and subsurface changes, such as the distribution of seasonal sea-ice cover, the appearance of invasive marine species, and changing water pH and temperatures can all have potentially dramatic influences on the distribution and abundance of desirable fish (Hannah and others, 2009). Since the 1970s, the Bering Sea has gradually shifted from a primarily cold Arctic marine ecosystem to a sub-Arctic system (Grebmeier and others, 2006); ocean and air temperatures have increased, and sea ice is less extensive in the southern Bering Sea (National Oceanic and Atmospheric Administration, 2010). Marine species composition has shifted in the southern portion of the Bering Sea, with dramatic increase in Walleye pollock, some increase in humpback and fin whales, and declines in Greenland halibut, snow crab, and fur seal (Newsome and others, 2007). Likewise, impacts continue to be observed for important subsistence mammals, such as walrus and polar bear, for which part or all of their life cycles depend in some way on the distribution and abundance of sea ice (Stirling and Parkinson, 2006; Laidre and others, 2008).

Shipping

Warming and other aspects of climate change have significant implications for marine access in the Alaska region. Likewise, climate change will likely have a major impact on both the types and amount of marine traffic. Although projected changes would bring economic opportunities to many cargo carriers, cruise operators, and some fishing fleets, they also will present sizeable infrastructure challenges and safety issues.

Loss of sea ice and changes in sea-ice character will undoubtedly be primary drivers of future shipping patterns in Alaska and the Arctic. Since the 1970s, the extent of Arctic sea ice has declined by roughly 10 percent (Shulski and Wendler, 2007). Decreasing Arctic sea ice extent also has been accompanied by pronounced thinning (Cavalieri and others, 2003), and a marked reduction in the age of the ice (Maslanik and others, 2011). Observations from Alaskan waters show similar trends, as well as an increasing distance from shore to ice-covered areas in summer and fall. Rates of future sea-ice change remain highly uncertain (Intergovernmental Panel on Climate Change, 2007), but longer term trends are expected to follow the direction of recent observations.

At present, much of the marine shipping activity in Alaska centers on ferry traffic, the transport of natural resources, and the delivery of general cargo and supplies to communities and resource extraction facilities. Offshore oil and gas operations are often supported by ships, and tugs and barges are critical for pollution response. Although much of this current activity is of regional scope, changing sea ice has the potential to bring a significant shift towards more long-distance and international traffic. As examples, reliable summer access to the Northwest Passage is expected in coming decades, and projections call for increased access to the Northern Sea Route (U.S. Department of Defense, 2011). In both cases, these changes would be accompanied by a much greater number of transits through the Bering Sea and other Arctic waters (Arctic Council, 2009), and the potential for increasing traffic through waters immediately adjacent to Alaska. At the same time, projected sea ice changes would likely lengthen the open season for many current routes.

Throughout Alaska and the Arctic, tourism in the form of traditional and adventure cruises is both common and on the rise. Between 2004 and 2007, for example, cruise ship traffic in the Arctic increased by 400 percent (Friends of the Earth, 2012). Continued reductions in sea ice would likely present further opportunities for development of the cruise industry through the opening of new routes and by extension of the current cruise season. Cruises centered on glacier viewing have a significant impact on Alaska's economy, and climate change is likely to impact these operations in a variety of ways. Continued loss of glaciers in some areas will likely necessitate a change in routes, with the potential for greater times at sea and/or longer open-water crossings in order to meet customer expectations. However, loss of glacial ice may provide greater access to areas where marine passage previously was limited or impossible.

In addition to changes in sea ice, many future projections call for changing storm frequency and intensity across the region (Intergovernmental Panel on Climate Change, 2007). Through a variety of mechanisms, increasing temperatures are likely to promote a northward shift in Pacific storm tracks (Yin, 2005; Salathé, 2006), while moisture from a more ice-free Arctic Ocean might increase storm intensity (Bengtsson and others, 2006). Impacts may be especially pronounced on the Bering Sea, where model-based studies indicate a strengthening of persistent low-pressure systems (Intergovernmental Panel on Climate Change, 2007). Although such changes would likely bring critical impacts to commercial fishing fleets, passenger ferry services, cruise liners and cargo shipping in open water, under these same scenarios the effects on near-shore operations and coastal infrastructure could be even greater. Even if the overall strength and/or frequency of storms stays the same, trends toward declining sea ice have the potential to amplify storm-related impacts on ports and other key components of shipping infrastructure (Wendler and others, 2010).

The Arctic Council's "Arctic Marine Shipping Assessment" (AMSA; Arctic Council, 2009), found that the potential for greater accessibility and resource development also would necessitate major investments in Alaska's marine-related infrastructure (see "Northern Sea Route"). As one example, the report pointed to the need for a much greater presence of icebreaking ships and ice-capable vessels to facilitate traffic in the early summer and late fall. The AMSA (Arctic Council, 2009) report and others (U.S. Department of Defense, 2011) also highlight the actions of other Arctic nations as potential responses to climate change in the Alaska region. In any case, climate change will likely require significant alterations to the various marine fleets and marine infrastructure operating in Alaska.

Department of Defense Operations in Alaska

The U.S. Department of Defense (DoD) recognizes that climate change presents increasing challenges for current and future operations, training, built infrastructure and natural resources on military lands. The DoD 2010 Quadrennial Defense Review (QDR; U.S. Department of Defense, 2010) included language that explicitly recognizes climate-change impacts for DoD. DoD operational readiness is contingent on access to land, air, and sea training and test space. Consequently, the 2010 QDR requires that "...the Department (of Defense) must complete a comprehensive assessment of all installations to assess the potential impacts of climate change on its missions and adapt as required (U.S. Department of Defense, 2010)."

Northern Sea Route

The Northern Sea Route (NSR), a shipping route traversing primarily Russian waters, is a viable and economic alternative to the Suez and Panama Canals. It is defined by Russian law as "a set of marine routes from the Kara Gate in the west to the Bering Strait in the east." Although the route is already an important national waterway, Russia seeks to capitalize changing conditions in the Arctic by transforming the NSR into a commercial shipping route of global importance capable of competing with more traditional routes in price, safety, and quality. The lack of pack ice in 2011 resulted in the longest navigational season on record for the NSR. Anticipating increases in cargo transport from 1.8 million tons in 2010 to 64 million tons by 2020, Russia is investing heavily in the Northern Sea Route by building 10 major rescue centers, and by pursuing national legislation that would, among other things, clarify tariffs for icebreaker assistance along the route (Pettersen, 2011). Russia also is in the process of building four diesel icebreakers and has plans to deploy the orbital monitoring system "Arktika" that will assist in vessel tracking and management.

Photograph by Sophie Webb, U.S. Geological Survey.

Warming temperatures and changes in precipitation will have a variety of implications for DoD installations throughout Alaska. The majority of built facilities and training lands are in south-central and interior Alaska, but more remote facilities are located along the coasts. The greatest implications will result from the retreat of sea ice and warming/thawing of permafrost soils. Vulnerabilities for DoD Alaska Region installations can be expressed through impacts in three broad categories: built infrastructure, operations and training, and management of natural resources.

Impacts on installation built infrastructure and equipment will differ depending on location within the State. Although higher temperatures will result in less heating requirements, they also will result in degradation of permafrost in interior Alaska and along much of the western and northern coasts, potentially damaging foundations, roads, pipelines, and communications structures. Many remote facilities (both active and inactive) along the Arctic and Western Alaska coast are of particular concern, where loss of sea ice is resulting in shoreline erosion rates of up to 100 ft annually (Kinner and others, 2009). Building and structure foundations, roads, and pipelines may require retrofits to protect their integrity due to increased active layer thickness of permafrost soils. Increased precipitation may come as rain in south-central Alaska but as more snow in the interior and coastal areas, resulting in increased snow loads on structures, potential design adaptations for existing buildings, and increased maintenance costs for snow removal.

Although impacts on air-based training and operations are expected to be minimal, land-based training will be affected mostly by changes in access to training areas. Many of the training areas in interior Alaska are utilized for winter training, when wetland areas and permafrost soils are frozen and snow covered. Access to some of these training lands is by ice bridges constructed in the winter over the Tanana and Delta Rivers. Increases in temperature and changes to permafrost will result in shorter durations of training access with some training areas becoming unusable. Increased drying conditions may result in some impact areas being unavailable for incendiary or pyrotechnic ordinances; live fire exercises also may be curtailed. Sea-based training in the Joint Pacific Alaska Range Complex (Gulf of Alaska) will be minimally affected but sea-based operations for the U.S. Coast Guard will likely be expanded as trans-Arctic and destinational shipping in the Arctic increases with the retreat of sea ice. The U.S. Coast Guard High Latitude Study included efforts to simulate Forward Operating Base activities in Barrow, Prudhoe Bay, and Nome in the summers of 2008 and 2009 (U.S. Government Accountability Office, 2011). These test operations indicated that the current suite of material (ships and aircraft) is not particularly suited to the shallow waters, long distances, and severe weather conditions experienced along the Arctic coast.

Environmental management of natural resources and regulatory compliance programs will be directly affected by temperature and precipitation trends, as well as the resulting change in permafrost and soil moisture. Federally listed species (endangered, at-risk, species of concern) have the most direct effect through restrictions on access to and utilization of training lands. Habitat transition or modification as a result of increased temperature, drought, altered hydrology, and alteration of fire regimes with climate change will complicate the ability of installations to maintain the status of federally listed species populations. Warmer temperatures may expand the northern limits of native and invasive species, resulting in habitat changes. Changes in permafrost and soil moisture may result in entire ecosystem shifts in interior Alaska, with permafrost supported wetlands draining and transitioning to a willow/scrub habitat.

The DoD is just beginning to develop and implement the necessary policy, guidance, technical capabilities and resources to effectively assess vulnerabilities, plan for, and adapt to potential climate change impacts. Development of technical support capabilities for DoD vulnerability assessment and adaptation planning have been initiated in Alaska through programs of the DoD Strategic Environmental Research and Development Program (SERDP; 2012). Three SERDP projects concerning the effects of climate change are currently (2012) being conducted in Alaska: determining the mechanistic links among fire, soils, permafrost, and vegetation succession; understanding permafrost hydrology, climate modeling, and ecosystem responses to change; and modeling permafrost, groundwater, and surface- water interactions.

Agriculture

The 1999 National Climate Assessment Alaska Technical Regional Report (Alaska Regional Assessment Group, 1999) mentioned the primary agricultural products and enterprises associated with agriculture in Alaska, and these have not changed much since then. The total area farmed in Alaska has decreased steadily in the last 30 years, from 1,287,000 acres in 1997 to 881,600 acres in 2007; total number of farms, however, increased from 383 to 686 during that period (U.S. Department of Agriculture, 2009). Mills (1994) calculated that Alaska has more than 101 million acres of potentially farmable land. The Alaska's Climate Change Strategy–Addressing Impacts in Alaska (State of Alaska, 2010) reported impacts on agriculture may be difficult to predict. Growing-degree days have increased by 20 percent, with potential benefits for agriculture (Arctic Climate Impact Assessment, 2004; Weller, 2005), although benefits of a longer growing season may be offset by the negative effects of decreased soil moisture (Karl and others, 2009a).

The positive and negative effects of climate change on agricultural products listed in 1999 are still valid today (see "Changing Climate Impacts on Agriculture") and were addressed in the State of Alaska Climate Reports (State of Alaska, 2010). Under a climate scenario of CO_2 doubling, Mills (1994) showed a significantly warmer and somewhat drier environment for northern Canada and Alaska and an increase in area of potentially arable soils. In Alaska specifically, Mills (1994) projected an increase of arable land from 47.6 to 98.1 million acres. In response to a changing agricultural situation in Alaska, the State of Alaska would support and expand sustainable agriculture production and marketing in Alaska under its NS-6 Sustainable Agriculture (State of Alaska, 2010).

The NS-6 report recommended four key actions to increase food security, to be led by the Alaska Division of Natural Resources Division of Agriculture:

1. Encourage community-based agriculture and practices that optimize the use of the land and resources available;

2. Research the magnitude and composition of food consumption in the State;

3. Research the sources of food supply and the risk associated with high reliance on imported foods; and

4. Develop, in cooperation with stakeholders, a strategic Alaska food policy to increase reliance on locally produced food sources through agriculture, seafood harvesting, and subsistence activities, including increased intrastate marketing of Alaska-grown products.

Changing Climate Impacts on Agriculture

Negative
- Changes in precipitation
- Changes in water balance (drying) due to permafrost degradation
- Increased erosion
- Increased insect infestation
- Greater potential for fires
- Increase in evapotranspiration could result in drought stress

Positive
- Potential new crops and animal husbandry
- Warmer temperature (extended plant range)
- Increased length of growing season

Potential Effects of a Changing Climate in Alaska

The current effects of a changing climate are well documented, but the potential future outcomes are, by definition, less certain and can only be hypothesized or modeled. Some projected and potential environmental properties that a changing (and predicted warming) climate may have on various aspects of Alaskan ecosystems, both environmentally and socially, are described below.

Marine Fisheries

In the North Pacific, forecasted warming trends, coupled with declining sea ice, raise concerns about the effects of climate change on harvestable fish and shellfish populations, habitat conditions, and ecosystem dynamics. In Alaska, the potential effects of climate-ocean variability on commercial fisheries are studied within very large marine ecosystems spanning the Gulf of Alaska, Aleutian Islands, and Bering Sea. One predicted outcome of climate change is that it will drive species ranges towards the poles (Parmesan and Yohe, 2003). This is of special concern in the northern Bering Sea and Arctic Ocean because resource assessments are inadequate for ecosystem-based management of fisheries (National Marine Fisheries Service, 1999; Pikitch and others, 2004; Fluharty, 2005; Francis and others, 2007; Marasco and others, 2007; Wilson and Ormseth, 2009). As a precautionary measure, a moratorium on commercial fishing in offshore waters in U.S. sectors of the Chukchi and Beaufort Seas was issued by the North Pacific Fishery Management Council (2009).

In the northern Bering Sea, concerns about fishing impacts on northwardly expanding populations of commercial resources has resulted in closures in bottom trawling in deep basin and slope areas and shelf waters in the Northern Bering Sea Research Area (North Pacific Fishery Management Council, 2011). These prohibitions reflect "risk-adverse" management actions in response to uncertainties associated with climatic effects on ecosystems, stock assessments, and changing fisheries (Stram and Evans, 2009; Witherell, 2009; Hollowed and others, 2011).

Current information indicates that the distribution and abundance of groundfish stocks off the coast of Alaska will be influenced by climate change (Mueter and others, 2007; North Pacific Fishery Management Council, 2010; Hollowed and others, 2011). Linkages between climate forcing and population processes may strongly influence the distribution and abundance of fish populations through changes in growth, survival, reproduction, or responses to changes at other trophic levels (Perry and others, 2005). Marine fish populations are governed by ecosystem processes, such as predation, competition, and environmental variability, and anthropogenic factors such as fishing. Variability in population rates of recruitment, natural mortality, growth, and catchability result from interactions among these processes, which are the dominant drivers of stock fluctuations (Maunder and Watters, 2003). Regional changes in ocean conditions (that is, weather; water temperature; hydrography, circulation and transport processes; nutrient dynamics; and chemistry) will interact with these processes and likely have the most profound effects on early life history stages subject to both bottom-up and top down controls (Mundy, 2005; Yatsu and others, 2008). Among the environmental changes, warming associated with climate change is seen as a particularly important threat to fish because it controls their environmental physiology and immune response and may result in large-scale shifts in host-pathogen relationships (Woodson and others, 2011).

Alaska's domestic groundfisheries represent the largest fishery by volume in the U.S. and target Pollock (*Theragra chalcogramma*), Pacific cod (*Gadus macrophalus*), sablefish (*Anoplopoma fimbria*), Atka mackerel (*Pleurogrammus monopterygius*), and numerous rockfish and flatfish; in the aggregate, these fisheries capture more than 130 species. More than 50 years of catch records illustrate regional trends in fisheries and population abundance and establish important baselines to assess possible climate effects (fig. 39). For example, as proxies for population abundance, they indicate large-scale changes in survival for some species and considerable interannual variability. Taken as a whole, the data show a system-wide regime shift in temperature and survival within the Gulf of Alaska, Aleutian Islands, and Bering Sea in the late 1970s and greater variability in the late 1980s (Mueter

and others, 2007). Population-level information is available for 50 harvested species, which represents a biological basis for integrated population modeling and the tools needed to assess, predict, and understand the effects of climate change.

Because salmon species spend most of their lives at sea, recent research has focused on climate impacts on early-marine and oceanic phases of their life cycles. The National Oceanic and Atmospheric Administration and others are investigating salmon abundance in Alaska with respect to large-scale atmospheric and ocean conditions affected by the Aleutian Low Pressure System (National Oceanic and Atmospheric Administration, 2012b). Recent trends in salmon production have been attributed to Pacific Decadal Ocean scale variability (Beamish and others, 1999; Hare and others, 1999), ocean temperature (Downton and Miller, 1998), and regional-scale sea surface temperatures (Mueter and others, 2002). As an example, stock productivity since the 1976 regime change in the North Pacific was estimated to be three times higher than that observed during the 1946–75 period (that is, Beamish and Bouillon, 1993; Mantua and others, 1997; Coronado and Hilborn, 1998). Ocean conditions, including storm events and upwelling in the Alaska Gyre, may have increased biological productivity, food availability, and survival of migrating salmon. Increase in survival was accompanied by a decrease in average salmon weight at maturity, 1975–1993, which was attributed to density dependence (Ishida and others, 1993; Bigler and others, 1996), sea-surface temperature (Ishida and others, 1995; Mueter and others, 2002; Hinch and others,

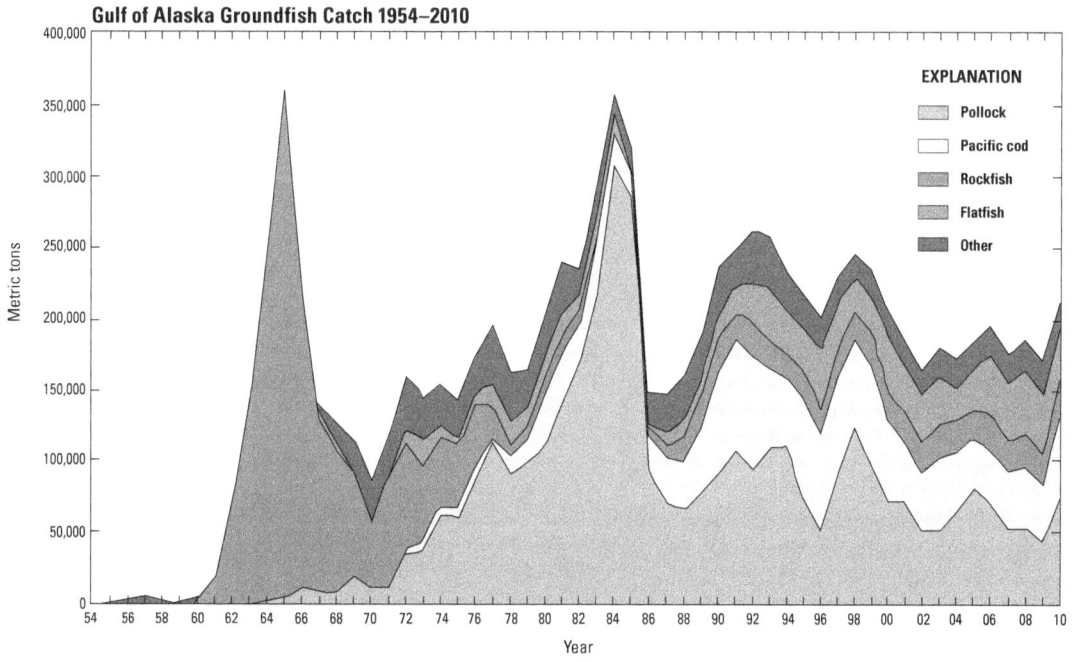

Figure 39.   Alaskan Groundfish Catch from 1954 to 2010 in the Gulf of Alaska. (Source: North Pacific Fishery Management Council, 2011.)

2005), and sea-surface salinity (Morita and Fukawaka, 2007). Exceptions to this decreasing trend include salmon from the Arctic-Yukon-Kuskokwim (AYK) management area (see Arctic-Yukon-Kuskokwim Sustainable Salmon Initiative, 2012). Current models are investigating salmon survival by integrating present-day climate and habitat variables (changes in precipitation, seasonal timing of runoff, temperatures, winds, and oceanic conditions) into population models. Preliminary analysis suggests that long-term sea-surface temperatures and plankton changes after 1996 are major contributors to the decline of AYK salmon populations, with suspected high mortalities associated with early-marine portions of the seaward migration (Myers and others, 2010).

Climate models for future greenhouse emissions for the northeast Pacific predict shifts in key processes (for example, precipitation, temperature, winds) and shifts in sea-surface temperatures and other conditions that, over time, may effect thermal ranges, seasonal quality of habitats (Azumaya and others, 2007), and movements of ocean-going salmon in the northeast Pacific and Arctic Ocean during the mid- to late 21st century (Randall and others, 2007; Abdul-Aziz and others, 2011). Welch and others (1998) examined how predicted ocean surface temperature changes would affect thermal habitat of sockeye salmon and found that the ocean area and migratory corridors suitable for sockeye could be greatly restricted and that their summer distribution might be spatially limited to feeding in the Bering and Othotsk Seas. In an expanded analysis, Abdul-Aziz and others (2011) projected thermal effects on winter and summer ocean habitats for all five species of Pacific salmon and steelhead. The spatial modeling predicted a decrease of 38 percent in winter habitats for sockeye (*Oncorhynchus nerka*), and decreases in summer habitats of 86 percent Chinook (*O. tshawytscha*), 45 percent for sockeye, 36 percent for steelhead (*O. mykiss*), 30 percent for coho (*O. kisutch*), 30 percent for pinks (*O. gorbuscha*), and 29 percent for chum (*O. keta*) by 2100. Projected habitat losses were largest in the Gulf of Alaska and in the western and central sub-Arctic North Pacific.

Subsistence economies, like those in Alaska, also may be affected, but it is unclear how social systems will respond to changes in the marine environment, because social and ecological couplings for remote Alaska communities, while evident, are non-linear and difficult to track (Huntington and others, 2009). Natural resource governance systems that delegate appropriate levels of decisionmaking authority with the resource users themselves (for example, fishermen) have been shown to help people cope with changing environments by coupling some degree of conservation responsibility with harvest rights (Loring and others, 2011). However, some approaches that achieve these effects in the short term, for example, Individual Fishing Quotas, may have societal ramifications that make them unsustainable in the long term (Carothers and others, 2010). Management agencies, such as the National Marine Fisheries Service, are already working to devise adaptive strategies for coping with changing fish stocks, habitat, and climate (Stram and Evans, 2009).

Terrestrial Landscapes

Broadly speaking, climate warming would affect a number of terrestrial landscapes in Alaska that are important to fish and wildlife through a variety of physical processes including thawing of permafrost and/or changes in permafrost dynamics; changes in snow and ice cover; changes in glacier dynamics; and altered hydrologic regimes (Hinzman and others, 2005; Martin and others, 2009; Chapin and others, 2010). In turn, key ecological responses are likely to incorporate one or more of the following factors, all of which are discussed elsewhere in this report:

- Changes in disturbance regimes, particularly fire and insect outbreaks;
- Lengthening of the growing season with concomitant changes in phenology and growth rates;
- Shifts in species distributions; and
- Introduction of novel species.

The effects of warming will be further mediated by any changes in precipitation (quantity, timing, or form), as well as the potential for changes in climatic extremes, such as droughts, storm events, or cold-air outbreaks (Intergovernmental Panel on Climate Change, 2007). Interactions among climate change, land use, management practices, and disturbance will ultimately determine the impacts to regional habitats (Jackson and others, 2009).

Potential habitat change in northern portions of Alaska has received a great deal of attention due to rapid rates of observed warming, and the possibility of "Arctic amplification" and other factors leading to relatively large amounts of future warming (Hinzman and others, 2005). As summarized by Martin and others (2009), changes in permafrost, alterations of local to regional-scale hydrology, and disturbance are likely to be the largest drivers of habitat change in this region. Significant changes in plant phenology, shrub and woody-plant dominance, and productivity also are expected (Sturm and others, 2001; Tape and others, 2006). Warming-induced extensions of the growing season and decreased soil moisture and permafrost stability are associated with increased abundance and extent of shrub cover (Tape and others, 2006) and increasing rates of coastal and stream erosion (Jones and others, 2009b; Tape and others, 2011).

On the Arctic Coastal plan, many projected habitat changes center around alterations to the numerous wetlands, ponds, and small lakes that characterize this area (Jorgenson and Shur, 2007; Martin and others, 2009), potentially affecting shore and water birds. Given the near certainty of future warming, ice wedges and other features of the underlying permafrost are likely to degrade substantially. Loss of subsurface ice would lead to a significant redistribution of water on the landscape. If this warming is combined with a wetter climate, then deeper lakes or low-lying basins are still likely to experience recharge; however, if future climate does not include enough precipitation to offset increased evapotranspiration,

Projected Effects of Climate Change on Polar Bears and Pacific Walrus

How polar bears and Pacific walrus will be affected by climate change later in this century has been a source of speculation and controversy. Polar bears are well-known for their occupation of sea ice for much of their annual cycle. Projections of sea ice loss by climate models are expected by most experts to have severe consequences to polar bears because studies documenting existing negative impacts to multiple polar bear populations are increasingly common (Stirling and others, 1999; Regehr and others, 2007, 2009; Rode and others, 2010, 2012). Negative consequences to polar bears as a result of loss of optimal sea ice habitat (Durner and others, 2009) are likely mediated by reduced access to their primary food source, ice seals. Growth rates of polar bear populations in the Southern Beaufort Sea (Point Barrow east) are linked to the amount of time sea ice is present over productive shallow waters, with growth rates declining in years of massive ice loss (Regehr and others, 2009). Simulations made with models that are based on this relation suggest the Southern Beaufort Sea polar bear population will decline to just a fraction of its current size by the end of this century (Hunter and others, 2010). Additional modeling suggests that if current patterns of sea ice loss continue, up to two-thirds of the world's current polar bear population will disappear by the end of this century (Amstrup and others, 2008). Results of model simulations also suggest, however, that greenhouse gas mitigation will foster persistence in the world's polar bears (Amstrup and others, 2010).

Projections on the impacts of a warming climate on Pacific walrus are more speculative because this species is very difficult to study. Jay and others (2011) developed a Bayesian network model to integrate potential effects of changing environmental conditions and anthropogenic stressors on the future status of the Pacific walrus population. Outcome probabilities through the century reflected a clear trend of worsening conditions for Pacific walrus. In the model, sea ice habitat and harvest levels had the greatest influence on future population outcomes.

Photograph by Dan Monson, U.S. Geological Survey.

desiccation of wetlands and ponds is likely (Smol and Douglas, 2007). Shifts or losses in wetland vegetation likely would occur, with drained basins serving as colonization sites for upland species. In larger lakes of the far north, shoreline erosion caused by increasing storms or thermokarst development could increase lake-surface area (Jorgenson and others, 2003), with accompanying consequences for lacustrine and lakeshore habitats.

In the Arctic uplands (that is, hills and mountains), areas dominated by loess and colluvium are thought to be highly sensitive to climate change because of the prevalence of ice-rich ground (Martin and others, 2009). On hillslopes and steeper terrain, increased thaw slumping is highly likely under a warming climate, and gullies also are likely to form in steep areas underlain by ice-rich soils. Fire has been historically uncommon in these areas, but events such as the 2007 Anaktuvuk River Fire point to the potential for large, habitat-altering disturbances in the Alaskan Arctic (Mack and others, 2011).

Riverine and floodplain habitats within northern Alaska are likely to show complex responses to future climate change depending on the balance of precipitation and evapotranspiration. Storm events, erosion, and sedimentation may influence the effect of warming on these habitats, with increased flooding potentially favoring more productive early-successional habitats. The prominence of underlying melting ice wedges (Martin and others, 2009) also may have an effect on these habitats, creating complex drainage systems that may move water away from these areas and creating opportunities for increased groundwater/surface-water interactions (Hinzman and others, 2005). In contrast, decreased stream discharge might lead to increased channel stability and shrub growth on the flood plain. Connectivity of oxbow lakes and other key bird and fish habitats would likely change.

Chapin and others (2010) identify four primary drivers of future habitat change in Alaska's interior and boreal forest zones:

1. Changes in soil moisture and hydrology associated with permafrost degradation and alterations to the regional water balance;

2. Changes in disturbance including fire, flooding, and insect or pathogen outbreaks;

3. Changes in the abundance or distribution of keystone species, such as white spruce, alder, and sphagnum moss; and

4. Interactions between a changing climate and human uses of the landscape.

Much of Alaska's boreal forest zone is characterized by discontinuous permafrost. Under a warming climate one would likely expect to see a continued loss of permafrost and attendant thermokarst formation. Depending on terrain, aspect, and surface vegetation, there may be a wide variety of habitat responses to permafrost degradation (Jorgensen and others, 2010). In lowlands and poorly drained sites, loss of ice-rich permafrost could lead to surface subsidence and conversion of forests to wetlands or ponds. In other areas, permafrost degradation would lead to pond or lake drying as the loss of underlying ice opens new pathways for water movement (Riordan and others, 2006). Climate-related changes in evapotranspiration and seasonal runoff also are likely to effect regional hydrology (Chapin and others, 2010).

Alaska's boreal regions are experiencing land-cover change, and a transforming wildfire regime is one of the primary causes (fig. 40). One-third of Alaska is covered by forest, and 90 percent of Alaska's forest is classified as boreal (Chapin and others, 2006). In the boreal forest, wildfires have dominated the disturbance regime of Alaska for approximately 6,000 years (Lynch and others, 2002). During recent decades, however, wildfire activity has accelerated. Change in activity has been attributed to "unusually" warmer and drier years likely caused by a directionally changing climate (Chapin and others, 2008). Previous climate-change assessments reported pronounced increases in extent, severity, and frequency of wildfire in Alaska over the past 60 years (National Assessment Synthesis Team, 2000; Kasischke and Turetsky, 2006; Karl and others, 2009a). During the first decade of the 2000s, the area burned and the number of large fires (>50,000 ha) were greater than any other decade since the 1940s (Kasischke and others, 2010; fig. 40), with the largest and third largest fire seasons on record occurring in 2004 and 2005, respectively.

Figure 40. The 2004 Boundary Fire, Alaska's largest wildfire season on record, which burned nearly 537,000 acres of forest in interior Alaska. Photograph by State of Alaska, Division of Forestry.

Black spruce forests, the dominant forest community type in the interior, historically burned during low-severity stand-replacing fires every 70–130 years (Johnstone and others, 2010a). In many areas underlain by permafrost, black spruce dominance is thought to be sustained by complex interactions with mosses on the forest floor that provide insulation and moisture for a spruce-rich seed bed (Johnstone and others, 2010a). Given predicted warming and permafrost thawing, however, this moss layer may dry becoming more susceptible to burning. If fires then expose mineral soils, the resulting successional trajectories often favor deciduous species, causing a shift towards deciduous-dominated forests in response to increased wildfire severity (Johnstone and Kasischke, 2005; Kasischke and Johnstone, 2005; Johnstone and Chapin, 2006) and reduction in fire-return interval (Johnstone and others, 2010a, 2010b; Bernhardt and others, 2011). On relatively dry sites (for example, south-facing slopes), there may be situations in which little or no post-fire tree regeneration would occur (Johnstone and others, 2010b). Thus, climate change is likely to affect the mechanisms that created the historical boreal zone landscape mosaic and bring widespread habitat change to these areas.

Rising temperatures have been shown to amplify insect outbreaks in Alaska (Arctic Climate Impact Assessment, 2004; Duffy and others, 2005; Juday and others, 2005; Balshi and others, 2009; Wolken and others, 2011). During the 1990s, south-central Alaska experienced the largest outbreak of spruce bark beetles in the world (Juday and others, 2005). This outbreak was associated with mild winters and warm temperatures that increased the over-winter survival of the spruce bark beetle and allowed the bark beetle to complete its life cycle in 1 year instead of them normal 2 years. This was superimposed on 9 years of drought stress between 1989 and 1997, which resulted in spruce trees that were too stressed to fight off the infestation. The forests of interior Alaska now may be threatened by an outbreak of spruce budworm, which generally erupts after hot, dry summers (Fleming and Volney, 1995). The spruce budworm has been a major insect pest in Canadian Forests, where it has erupted approximately every 30 years (Kurz and Apps, 1999), but was not able to reproduce in interior Alaska prior to 1990 (Juday and others, 2005).

In southeast Alaska, climate warming has affected forest ecosystems primarily through effects on the form of precipitation (that is, snow versus rain). For the past 100 years, the culturally and economically important yellowcedar has been dying in portions of the region (Hennon and others, 2006). The onset of decline in yellow cedar (*Callitropsis nootkatensis*) in 1880 (Hennon and others, 1990), with tree mortality rates of about 70 percent (D'Amore and Hennon, 2006) in this region, is attributed to warmer winters and reduced snow, combined with early spring freezing events (Hennon and others, 2006; Beier and others, 2008; Schaberg and others, 2011). The decline in yellowcedar has many societal consequences, as it is the highest valued commercial timber species exported from the region (Robertson and

Brooks, 2001). Native Alaskans also value this tree for ceremonial carvings; documented subsistence uses include fuel, clothing, baskets, bows, tea and medicine (Schroeder and Kookesh, 1990; Pojar and MacKinnon, 1994).

Invasive Species

Invasive species are defined under Executive Order 13112 (Code of Federal Regulations, Executive Order 13112, 1999) as species that are present in a particular ecosystem due to an intentional or unintentional escape, release, dissemination or placement into that ecosystem as a result of human activity (that is, "introduced" or "non-native" species), and whose introduction does or is likely to cause economic or environmental harm or harm to human health. Thus not all non-native species are considered invasive. However, some non-native species considered to pose no invasive threat at the time of introduction may exhibit explosive population growth and lead to invasive impacts long after their initial establishment in a new environment despite initially being considered benign (Sakai and others, 2001).

As climate change may potentially alter Alaska ecosystems and enable greater human activity, biological invasion in Alaska and across the Arctic is likely to increase. Arctic terrestrial ecosystems may be predisposed to invasion because many invasive plants are adapted to open disturbed areas (Hierro and others, 2006). If fire frequency and severity increase with climate change (Hu and others, 2010b), Arctic ecosystems may become more susceptible to further invasions. Areas of human disturbance and those located along pathways of human activity (for example, shipping and road corridors) are the most likely sites of further invasion into Alaska habitats. For example, Conn and others (2008) noted the susceptibility of gravel-rich river corridors to white sweet clover invasion dispersal from bridge crossings.

Carlson and Shephard (2007) describe Alaska as being in a "unique and advantageous position" with regard to the establishment of non-native plants because "the majority of land has not been impacted by human development, and non-native plants are still largely concentrated in high-use areas." Indeed, to date there are many fewer invasive terrestrial plants known from Alaska and Arctic Alaska in particular, than in other altered and invaded ecosystems of lower latitudes. In part this may reflect simple absence, but also may reflect a lack of regular monitoring. The U.S. Department of the Interior, Bureau of Land Management (2012) summarizes the status of invasive plants in the National Petroleum Reserve in Alaska (NPR-A) by saying that "little is known about non-native, invasive, plant species in the NPR-A." They point out that the Alaska Exotic Plants Information Clearinghouse (University of Alaska Anchorage, Alaska Natural Heritage Program, 2012a), an extensive database of invasive plant information for Alaska, includes very few survey data for the area north of the Brooks Range crest. They did note that a survey of the Dalton Highway (which leads to the

North Slope oil fields) "detected 28 species of non-native, invasive plants" north of the Yukon River, but only two (foxtail barley, *Hordeum jubatum*; and common dandelion, *Taraxacum officinale*) were found north of the Brooks Range. Importantly, the U.S. Department of the Interior, Bureau of Land Management (2012) noted that "highways such as the Dalton, rivers, and trails provide corridors" for the movement of invasive plants into un-invaded areas and that "equipment and vehicles used for exploration and construction" may act as vectors of spread from these corridors.

Carlson and Shephard (2007) nonetheless provide evidence of an accelerating rate of spread for non-native plants in Alaska and attribute this primarily to the increase in human population and associated "ground disturbing activities" such as "oil development, agriculture, housing, and roads." However, even in Alaska, invasive plants are not limited to disturbed sites. Carlson and Lapina (2004) had noted an increase in the movement of non-native species "off the anthropogenic footprint and into more intact ecosystems."

Dukes and Mooney (1999) noted that in general they "expect most aspects of global change to favor invasive alien species and thus exacerbate the impacts of invasions on ecosystems." In Alaska, there have been few studies specifically linking the occurrence of individual invaders with climate-induced changes.

With the majority of goods shipped to interior and northern Alaska via ports in south-central Alaska, invasive plant species likely will become an increasingly important risk factor. Several invasive plant species in Alaska spread aggressively into burned areas [for example, Siberian peashrub (*Caragana arborescens*), Narrowleaf hawksbeard (*Crepis tectorum*), and White sweetclover (*Melilotus alba*); Lapina and Carlson, 2004; Cortes-Burns and others, 2008], and as a result could increase with the increase in wild fire potential in this region.

In marine environments, Hanson and Sytsma (2007) estimated that Alaska waters are not currently (2012) at high risk of invasion from Chinese mitten crabs (*Eriocheir sinensis*); if water temperatures rise due to climate change, however, many Alaska estuaries would be at risk. In an analysis of the risks associated with the pathogen that causes whirling disease in salmonids (*Myxobolus cerebralis*), Arsan (2006) suggests that the risk of parasite dissemination in Alaska will vary with conditions that affect parasite development, such as climate change. Hines and others (2004) found that some Alaska coastal waters are already at risk of invasion by the European green crab (*Carcinus maenas*), and de Rivera and others (2007) suggested that several marine invasive species, including the European green crab, had the potential to expand to sub-Arctic and Arctic waters even under moderate climate change scenarios. Similarly, Ruiz and Hewitt (2009) concluded that "environmental changes may greatly increase invasion opportunity at high northern latitudes due to shipping, mineral exploration, shoreline development, and other human responses."

Stachowicz and others (2002) demonstrated the ability for climate change to directly increase invasion by the marine tunicate *Botrylloides violaceus*. This same species is known to have invaded Alaska waters (Ruiz and others, 2006), and is one of a suite of tunicate species, along with the European green crab, that are regularly sampled for by a collaboratively supported network of coastal Alaska communities from Ketchikan to Barrow. Another highly invasive tunicate, *Didemnum vexillum*, was verified in Alaska waters for the first time in the summer of 2010 and is now the subject of another collaboration of Federal, State, Tribal, university, and local entities in an effort to halt its further spread from the Sitka area to other Alaska waters (Smithsonian Environmental Research Center, 2010).

Another study found that the rate of marine invasion is increasing; that most reported invasions are by crustaceans and mollusks; and, importantly, that most invasions have resulted from shipping (Ruiz and others, 2000). The external hull and ballast tanks of vessels operating even in ice-covered waters can support a wide variety of non-native marine organisms (Lewis and others, 2003, 2004). Given the findings of the recent analysis of current Arctic shipping and the potential for climate change to expand such shipping (Arctic Council, 2009), this has relevance for future marine invasive risks to Alaska waters.

Tourism

Climate change is expected to present both opportunities and challenges to Alaska's tourism industry. Drawing on Arctic Council data of observed and projected decreases in Arctic sea ice, the Arctic Marine Shipping Assessment Report states, "Arctic marine tourism's most likely future is that larger numbers of tourists, traveling aboard increased numbers of ships of all types, will be spending more time at more locations" (Arctic Council, 2009). Assuming this occurs, coastal regions of Alaska may have opportunities for increased tourism development. However, some researchers warn that the changing sea ice regime may in fact create more hazardous ice conditions, resulting in a negative effect on cruise tourism in certain parts of the Arctic (Stewart and others, 2007).

Longer and warmer Alaskan summers will help increase the number of tourists coming to Alaska through extended operating seasons, as envisioned by the Alaska Climate Impact Assessment Commission (the Commission; 2008). The Commission suggests that wildlife viewing—a major tourism attraction in the State—may be enhanced by the positive effect of shorter and less severe winters on wildlife. In contrast, the Commission also puts forth the contrasting idea that winter tourism in the State may be negatively affected by climate change, as more unpredictable winter weather has already necessitated the cancellation or relocation of dog sled races, skiing events, and other winter activities. The Commission concludes that drier summers have led to increased wildfires, creating smoke that detracts from tourists' experiences. They also cite climate-change induced glacial shrinkage as a potential threat to the sightseeing cruises that dominate tourist activity in the southeast region of the State.

The likely mixed effects of climate change on Alaskan tourism are corroborated by the results of a study that presents a quantitatively modeled tourism climate index for two tourist destinations in the State: King Salmon and Anchorage (Yu and others, 2009). The results show that climate change will likely extend the summer sightseeing season at King Salmon but shorten the total time for skiing each winter in Anchorage.

Other studies indicate that as species migrate in response to climate variability and change, specific areas may no longer be able to support the flora and fauna that now reside there (U.S. Geological Survey, 2006; U.S. Environmental Protection Agency, 2011). Such ecological changes could negatively affect the independent nature guide industry that relies on the current flora and fauna profile to draw tourists to certain areas. Projected increases in the costs of maintaining public infrastructure caused by thawing permafrost (Larsen and others, 2007a) also stand to negatively affect aspects of the tourism industry that rely on harbors, roads, airports, and water and sewer systems. In addition to affecting Alaska's natural attractions and infrastructure, climate change is thought to be influencing the destination choices of tourists. The new trend is known variously as last-chance tourism, climate change sightseeing, or the tourism of doom (Lemelin and others, 2010; Schlichter, 2011). A changing Alaskan landscape may bring tourists to new locations, either because a certain attraction may not be there in the future (for example, coastal communities, polar bears, calving glaciers) or because it is thought to be where the effects of a warming climate will first be noticed and most dramatic (Rosen, 2007).

Summer tourism in Alaska (consisting largely of cruise ship visitors and accounting for the vast majority of annual visitor numbers and spending) is expected to experience a net benefit from climate change because of longer operating seasons. However, wildfires, changing ecosystems, shrinking glaciers, and degraded public infrastructure may pose challenges to this sector. Winter tourism (primarily snow- and ice-dependent activities that account for a minor but important percentage of annual visitor numbers and spending) is expected to absorb an overall negative effect from climate change as winters become warmer and shorter in the State.

Permafrost

A warming climate is predicted to have a number of potential impacts on permafrost and related ecosystem processes. Feedbacks to natural systems include the increase of hydrological connectivity from development of taliks (unfrozen layers) in the discontinuous permafrost zone, resulting in increased loss of lakes owing to subsurface drainage (Yoshikawa and Hinzman, 2003; Riordan and others, 2006) and increased exchange between surface water and groundwater and their included nutrients and contaminants (White and others, 2007; Alessa and others, 2008; Rowland and others, 2011). In the continuous permafrost zone,

hydrological changes may differ regionally, with increased lateral lake drainage in some regions due to thawing of ice-rich surface layers (Jones and others, 2011) and the formation of new ponds and lakes in other regions through the process of thermokarst (Smith and others, 2005). Increased input of nutrients and particulate matter into lakes and streams from thaw slumps and active layer detachment slides has been observed in Alaska and Northwest Canada (Kokelj and others, 2005; Bowden and others, 2008; Rowland and others, 2010), with effects ranging from fertilizing to clogging of downstream water bodies.

Globally significant impacts are expected to result from release of soil organic carbon stored in permafrost (Schuur and others, 2011). Substantial amounts of organic carbon, currently about twice the amount of carbon present in the atmosphere, have been stored for thousands of years in permafrost regions due to very slow or negligible decomposition of plant organic matter under below freezing temperatures (Ping and others, 2008; Tarnocai and others, 2009). If thawed, this soil organic carbon would decompose and be released as the greenhouse gases carbon dioxide or methane (Walter and others, 2007; Schuur and others, 2009), or leaked as dissolved organic carbon and particulate organic carbon into lakes, streams, and the sea (O'Donnell and others, 2010; Ping and others, 2011). Although current greenhouse gas fluxes from permafrost regions are small compared to anthropogenic emissions, this natural source may become much stronger due to permafrost thaw in the near future, causing accelerating feedbacks in the climate system. The release of only a fraction of the permafrost-stored soil carbon to the atmosphere would significantly increase atmospheric greenhouse gas contents. Mechanisms that result in such soil carbon release include press disturbances,[6] such as gradual but widespread top-down permafrost thawing and pulse disturbances such as rapid but local thermokarst, thermo-erosion, and wildfires (Grosse and others, 2011).

Oil, Gas, and Mining

Oil, gas, and mining are three of the leading sectors in the Alaska economy, generating more than 85 percent of State revenue from royalties and taxes as well as significant employment. The oil and gas industry is estimated to have the greatest potential for substantial economic growth in the Arctic. To support increased economic activity, ports, infrastructure, and other facilities are expected to be developed as warming temperatures result in longer seasonal access. This may bring increased ship traffic and a greater human presence, not only creating job and business opportunities, but also requiring investments to ensure that essential government functions, such as safety, security, and environmental protection, are provided.

---

[6]A disturbance that occurs as a gradual or cumulative pressure on a system is referred to as a "press" disturbance while a relatively discrete event in time is referred to as a "pulse" disturbance.

Temperature Trends in the National Petroleum Reserve

Recently compiled observations show winter air temperatures have been cooling across large portions of the Arctic, predominantly Scandinavia and Eurasia, over the last two decades (Cohen and others, 2012). USGS data indicate that this has not been the case in the National Petroleum Reserve—Alaska (NPR-A) in the central part of Alaska's Arctic. Rather, air temperatures have been warming during all seasons in the NPR-A over at least the last decade. The warming has been strongest during the winter and weakest during the summer, consistent with climate models. The spatial variability of the warming trends is most pronounced during the snow-free season (summer), as expected, since local meteorological processes tend to dominate at that time. Ground temperature trends generally reflect air temperature trends. The plots below show the temperature trends for five stations spanning the NPR-A.

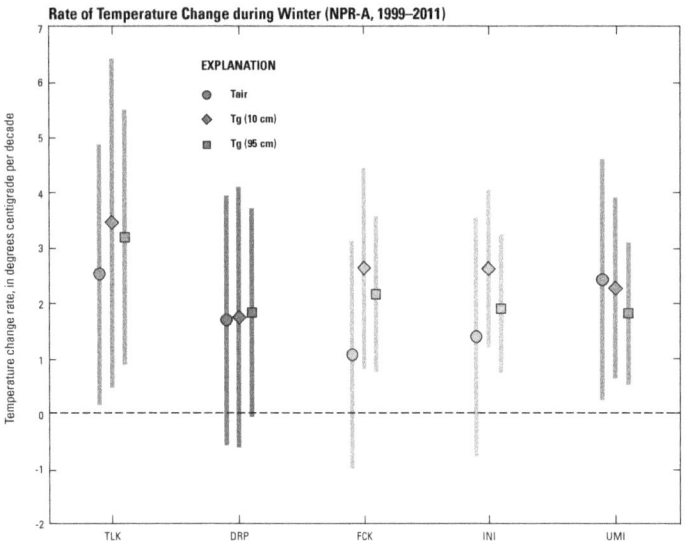

Rate of temperature change in the National Petroleum Reserve–Alaska during the winter, 1999–2011. Tair—temperature of air 3 m above ground; Tg (10 cm)—temperature of ground 10 cm below the surface; Tg (95 cm)—temperature of ground 95 cm below the surface. Point estimates of the warming trends are shown by symbols; the vertical lines show 90-percent confidence intervals. TLK (Tunalik, west coast), DRP (Drew Point, north coast), FCK (Fish Creek, north coastal plain), INI (Inigok, central coastal plain), UMI (Umiat, foothills).

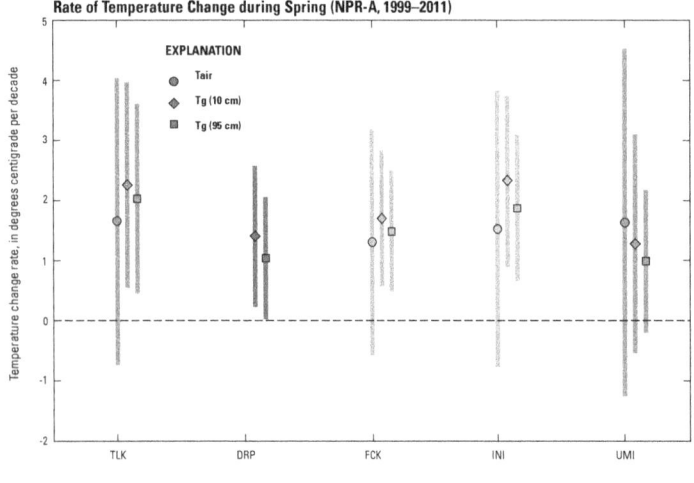

Rate of temperature change in the National Petroleum Reserve–Alaska during the spring (1999–2011). Tair—temperature of air 3 m above ground; Tg (10 cm)—temperature of ground 10 cm below the surface; Tg (95 cm)—temperature of ground 95 cm below the surface. Point estimates of the warming trends are shown by symbols; the vertical lines show 90-percent confidence intervals. TLK (Tunalik, west coast), DRP (Drew Point, north coast), FCK (Fish Creek, north coastal plain), INI (Inigok, central coastal plain), UMI (Umiat, foothills).

The oil and gas industry also may experience climate-change effects in Alaska that are not apparent in other regions of the United States. Potential drying of the landscape may cause a reduction in the availability of water used for snow travel and ice roads. As one example, the number of days per year in which travel on the tundra is allowed under Alaska Department of Natural Resources standards has decreased from more than 200 to about 100 days in the past 30 years. This results in a 50-percent decrease in days that oil and gas exploration and extraction equipment can be used (Karl and others, 2009a).

Alaska's economic future could potentially depend largely on maintaining robust oil and gas production because of reserves that exist in Alaska along the Beaufort Sea coast and in the Mackenzie River/Beaufort Sea area of Canada. Although climate-change impacts on this industry have been minor thus far, both negative and positive effects may likely occur in the future. For example, offshore oil exploration and production is likely to benefit from less extensive and thinner sea ice (Weller, 2005) and more jobs and improved medical care and schools (National Research Council, 2003). These economic benefits, however, have been accompanied by environmental and social consequences, including effects of the roads, infrastructure and activities of oil exploration and production on the terrain, plants, animals and peoples of the North Slope and the adjacent marine environment (National Research Council, 2003). Conversely, ice roads, now used widely for access to on- and off-shore activities and facilities, are likely to be less safe and usable for shorter periods (Weller, 2005), and continued warming will further impair transport by shortening the seasonal use of ice roads (Arctic Climate Impact Assessment, 2004).

With thawing permafrost, decreased sea ice extent and changing weather patterns, oil and gas operations may see impacts both onshore and offshore, such as impacts to infrastructure (for example pipelines, ice roads, and waste pits), exploration and production facilities (such as reduced efficiency of gas compression and reinjection), and shorter and warmer winters have already resulted in reduced operation windows for exploration and development. Engineering focused on proactively addressing challenges of the changing climate may be essential for this sector to remain viable. For example, the requirement to run onshore seismic exploration lines during the winter when the local northern environment is frozen and covered with snow, and therefore more resilient to vehicle traffic. Permitting for onshore exploration also is now normally conducted during the winter when stable man-made ice pads can be constructed and the movement of heavy drilling equipment across the onshore and offshore areas is more environmentally sound (Clow and others, 2011).

Alaska's other major economic subsurface resource is minerals, and mining, like oil and gas exploration, relies on engineered infrastructure that also may be potentially affected by climate and weather. Thawing permafrost and other climate-related changes could threaten the stability of crucial production and processing infrastructure, including transportation. Thawing is projected to accelerate under future warming, with as much as the top 10–30 ft of discontinuous permafrost thawing by 2100 (National Assessment Synthesis Team, 2000; Romanovsky and others, 2007). This may create special engineering challenges to existing and planned infrastructure and transportation (Alaska Department of Environmental Conservation, 2012).

Infrastructure

The thawing of permafrost will have profound effects on foundations and structures, and any time permafrost is disturbed for infrastructure construction, there is potential for thaw (Khrustalev, 2001; Nelson and others, 2001; Romanovsky and Osterkamp, 2001; U.S. Arctic Research Commission Permafrost Task Force, 2003; Arctic Climate Impact Assessment, 2005). Climate change and its associated warming add an additional layer of complexity and cost to reduce or mitigate infrastructure damage or settlement. Existing infrastructure may be further destabilized, requiring additional maintenance, rebuilding, and reinvestment (U.S. Arctic Research Commission Permafrost Task Force, 2003; Arctic Climate Impact Assessment, 2004). Direct impacts include damages to buildings, roads, and pipelines, and potential hazards caused by uneven ground surface settlement related to thawing of ground ice, resulting in additional maintenance, mitigation, adaptation, and/or relocation costs (see below). Although a majority of the population of Alaska resides in areas underlain by sporadic or less than 10 percent permafrost, many communities are still located in areas vulnerable to permafrost degradation; most of the State's major roads also are subject to the effects of thawing permafrost (fig. 41).

Warming also will accelerate the erosion of shorelines and riverbanks, threatening the infrastructure located in these areas (U.S. Arctic Research Commission Permafrost Task Force, 2003). Some villages or facilities located on riverbanks or exposed coastlines are facing major problems with erosion (U.S. Arctic Research Commission Permafrost Task Force, 2003; Arctic Climate Impact Assessment, 2004; Bronen, 2011), and several villages in Alaska have lost buildings to the sea (Callaway and others, 1999). The U.S. Army Corps of Engineers reports the villages of Shishmaref, Kivalina, and Newtok in western Alaska will need to be moved, and relocation will have to take into account current and changing near-future permafrost conditions at target sites. Based on permafrost model projections (Marchenko and others, 2008), the majority of northern Alaskan communities will be affected by permafrost thaw by 2100.

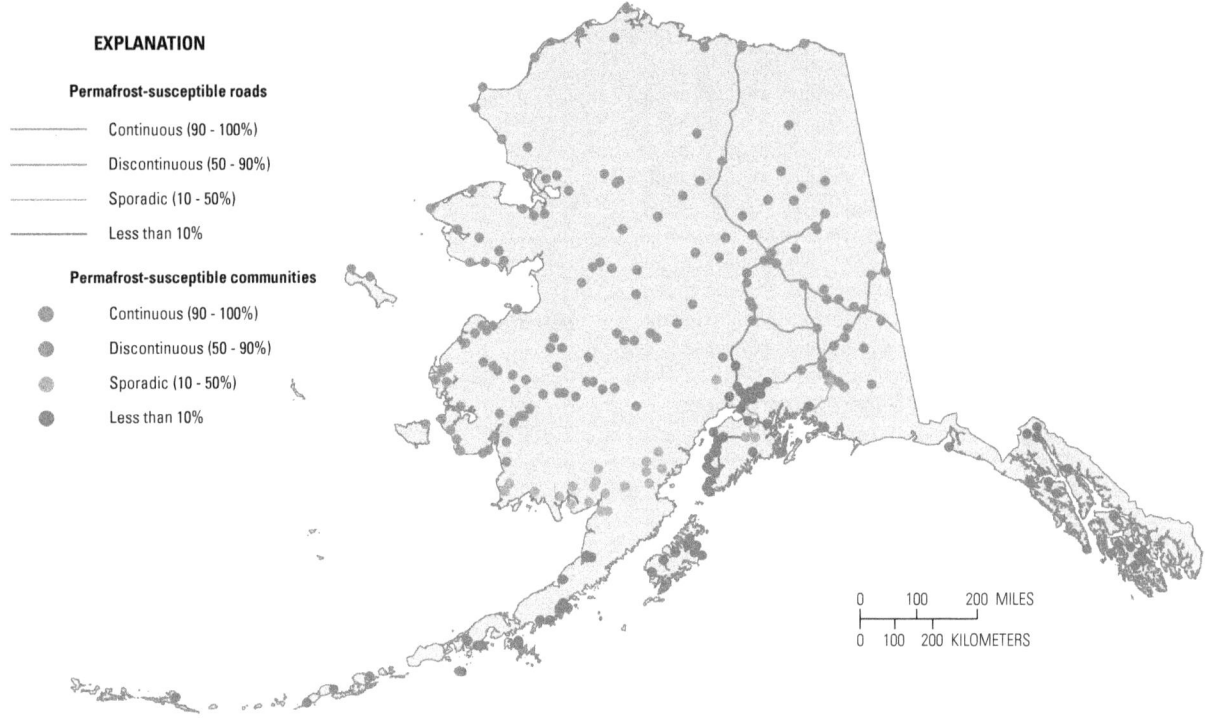

Summary of Alaska Highways Susceptible to Permafrost

Permafrost extent	Road distance, in miles
Continuous (90 - 100%)	456
Discontinuous (50 - 90%)	1,2111
Sporadic (10 - 50%)	189
Less than 10 %	281

Summary of Alaska Communities Susceptible to Permafrost

Permafrost Extent	Total Communities	Population
Continuous (90 - 100%)	87	40,811
Discontinuous (50 - 90%)	79	47,140
Sporadic (10 - 50%)	26	5,235
Less than 10 %	129	396,821

Figure 41. Communities and major roads in Alaska susceptible to the effects of thawing permafrost (figure adopted from U.S. Arctic Research Commission Permafrost Task Force, 2003).

Economic Impacts

Climate change will create both costs and potential benefits to Alaska's economy. A number of sources cite possible climate-related effects that could directly affect Alaska's economy including: expanded marine shipping and access to offshore minerals; declining food security; human health concerns; effects on ecosystems, wildlife, fisheries, and tourism; disrupted onshore transportation systems; and damage to community infrastructure from increasing fire activity, erosion and thawing permafrost (Alaska Regional Assessment Group, 1999; U.S. Arctic Research Commission Permafrost Task Force, 2003; Arctic Climate Impact Assessment, 2005; Hinzman and others, 2005; Trainor and others, 2009; Yu and others, 2009; Cherry and others, 2010; State of Alaska, 2010; U.S. Arctic Research Commission, 2010). To date, no known research has been conducted to quantify the possible economic benefits of climate change to Alaska. A limited number of studies have attempted to estimate the magnitude of climate-related costs, with researchers focusing primarily on the financial risk to Alaska's infrastructure.

Table 7.   Estimated protection and relocation costs for three Alaska communities.

[Source: U.S. Army Corps of Engineers (2006)]

Community	Costs of initial erosion protection ($millions)	Costs to relocate ($millions)	How long until relocation needed?
Kivalina	$15	$95–125	10–15 years
Newtok	$90	$80–130	10–15 years
Shishmaref	$16	$100–200	10–15 years
Totals	$121	$275–455	10–15 years

A 2006 report by the Alaska branch of the U.S. Army Corp of Engineers estimated relocation costs for villages affected by flooding and coastal erosion in western Alaska and that the communities of Shishmaref, Kivalina, and Newtok must be moved in the next decade to avoid catastrophic losses. Estimates for relocating those villages are as much as $455 million (U.S. General Accounting Office, 2003; U.S. Army Corps of Engineers, 2006; see table 7).

In 1999, Alaska civil engineering and planning experts speculated that the costs of dealing with infrastructure affected by climate change could exceed the budgets of many of the agencies responsible for their upkeep. Cole and others (1999) indicated that yearly costs for damages due to global climate change for the State of Alaska could be as high as $35 million, which is similar to the State and Federal costs for firefighting each year, and represents a sizeable fraction of the State's capital projects budgets (or about equal to the budgets of the Department of Fish and Game at $34 million and the Department of Natural Resources at $40 million). In 2007, a preliminary analysis found that climate change could add $3.6 to $6.1 billion—representing 10–20 percent increase above normal wear and tear—to future costs for public infrastructure from 2007 to 2030 (fig. 42). These estimates took into account different possible levels of climate change and assume government agencies partially offset the level of risk by strategically adapting infrastructure to changing conditions (Larsen and others, 2008; Chinowsky and others, 2010). However, subsequent analyses by some of the principal researchers involved in this study found that a number of factors may have contributed to a systematic underestimate of both the dollar amount of infrastructure at risk and the statistical uncertainty of their original results (Foster and Goldsmith, 2008; U.S. Arctic Research Commission, 2010). Additional risk to Alaska's private infrastructure also is likely, but there has been no effort to date to systematically quantify this vulnerability.

Arctic Climate Change—A New Normal?

The Arctic is showing large visible changes over the last decade and many of the shifts are indicators of major regional and global feedback processes (Kattsov and others, 2010). Of principal importance is "Arctic Amplification," whereby surface temperatures in the Arctic are increasing faster than elsewhere in the world (fig. 43). Further, changes in the Arctic are occurring faster than indicated by results of simulations made with coupled air-sea-ice climate models (Stroeve and others, 2007).

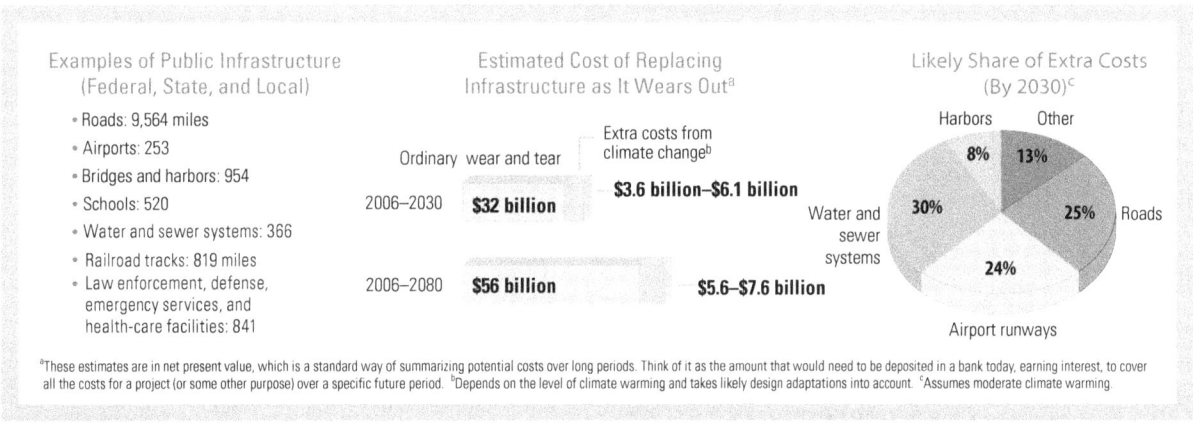

Figure 42.   How much climate change might add to future costs for public infrastructure in Alaska. (Source: Larsen and Goldsmith, 2007).

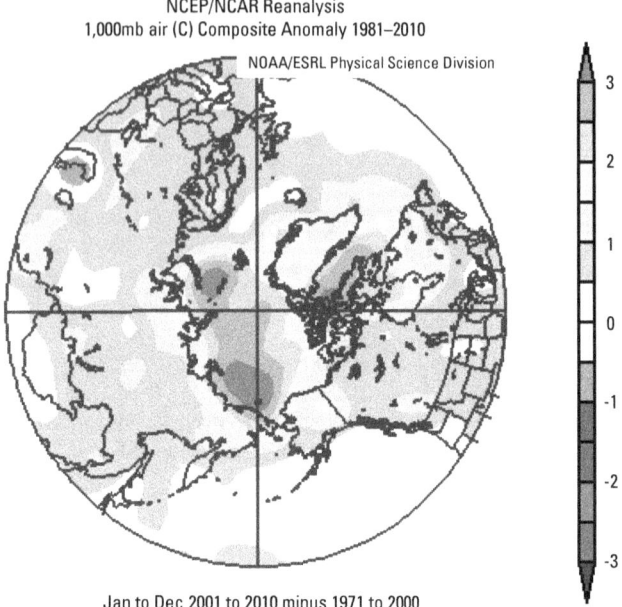

NCEP/NCAR Reanalysis
1,000mb air (C) Composite Anomaly 1981–2010

NOAA/ESRL Physical Science Division

Jan to Dec 2001 to 2010 minus 1971 to 2000

Figure 43. Near-surface air temperature anomaly multi-year composite for 2001–2010. Anomalies are relative to the 1971–2000 mean and show a strong Arctic amplification of recent temperature trends. Generated online at National Oceanic and Atmospheric Administration (NOAA) Earth System Research Laboratory (ESRL), Physical Sciences Division, website, accessed August 10, 2012, at http://www.esrl.noaa.gov/psd/. NCEP: National Centers for Environmental Protection; NCAR: National Center for Atmospheric Research; mb, millibar.

The rate of decline of Arctic sea ice thickness and September sea-ice extent has increased considerably during the first decade of the 21st century (Maslanik and others, 2007; Nghiem and others, 2007; Comiso and Nishio, 2008; Deser and Teng, 2008; Alekseev and others, 2009). By September 2007, the area of sea ice had declined to 37 percent of its extent during the period 1979–2000. Although at the time, it was unclear whether the record minimum extent of ice in 2007 was an extreme outlier, every year since then (2008–2011) has had a smaller September sea ice extent than the years before 2007, with 2011 being second lowest compared with 2007 (fig. 44). In addition, the amount of old, thick multi-year sea ice in the Arctic also has decreased by 42 percent from 2004 through 2008 (Giles and others, 2008; Kwok and Untersteiner, 2011) and the sea ice has become more mobile (Gascard and others, 2008). Thus, the Arctic may be moving toward a new state in which it is dominated by first year sea ice processes, and will lose some of the long term, more stable dynamics associated with old, thick sea ice.

Over the last 5 years, evidence has continued to accumulate from a range of observational studies that systematic changes are occurring in the Arctic. Persistent trends in many Arctic variables, including sea-ice extent, the timing of spring snow melt, increased shrubbiness in tundra regions, changes in permafrost, increased area coverage of forest fires, increased ocean temperatures, changes in ecosystems, as well as Arctic-wide increases in air temperatures, can no longer be associated solely with the dominant climate variability patterns, such as the Arctic Oscillation or Pacific North American pattern (Quadrelli and Wallace, 2004; Vorosmarty and others, 2008; Overland, 2009).

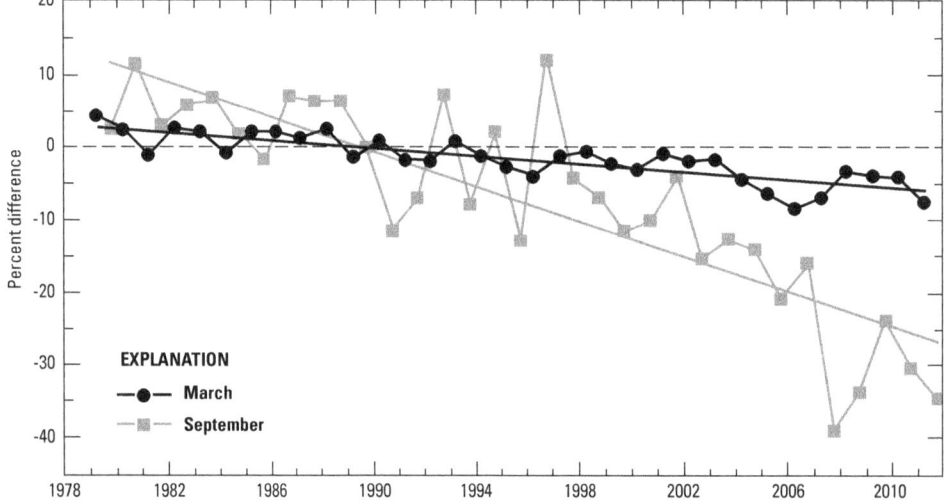

Figure 44. Time series of the percentage difference in Arctic sea ice extent in March (the month of maximum ice extent) and September (the month of minimum ice extent) relative to the mean values for the period 1979–2000. Based on a least squares linear regression for the period 1979–2011, the rate of decrease for the extent of ice in March and September is -2.7 and -12.0 percent per decade, respectively.

New Science Leadership on the Alaskan Landscape

Beginning in the early part of this decade, the idea of climate change and its potential effects on the ocean, land, and human environment became more prominent in the minds of Alaska's policymakers, land and resource managers, and health practitioners. As a result, a number of fact finding and science activities were initiated to provide information and science leadership to State and Federal organizations to discuss and initiate actions that would allow more informed choices on management, adaptation, and/or mitigation of climate impacts. This section describes those activities.

Alaska Climate Change Executive Roundtable

The Alaska Climate Change Executive Roundtable (ACCER) was established jointly by the U.S. Fish and Wildlife Service and the U.S. Geological Survey in 2007. It comprises both Federal and State senior level agency executives from throughout Alaska and an alliance of other members, each with responsibilities and capacities for addressing climate change or its impacts on Alaska's natural and cultural resources. Membership also includes Alaska Native leaders, who provide insight and perspective to assist in understanding climate's impact on Alaska Native communities across the State. The ACCER agency programs variously involve research, management, regulatory and service-oriented activities, and are responsive to a variety of customers and constituencies. Although there are differences in the programs, ACCER members agree that there are opportunities to exchange information, share best practices, leverage resources, and coordinate activities. This is particularly important in addressing large-scale conservation and management concerns in the face of climate change that may exceed the abilities and resources of any one agency and require a high level of collaboration across Alaska. Thus, the ACCER establishes an effective working environment to facilitate identification of shared needs and coordination of efforts among agencies and initiatives.

The overall mission of the ACCER is to promote a collaborative effort in advancing the knowledge of climate change variables as it relates to support each member's responsibility for effective adaptation and mitigation strategies and management responsibilities through information sharing and coordination among partners of existing and new capacities. Specific goals of the ACCER are:

1. To identify climate change-related issues of common concern. The ACCER forum allows Members to gain a greater appreciation of challenges facing Member agencies, share lessons and capabilities, collaborate on the development of strategic approaches, and harmonize plans and processes;

2. To facilitate collaborative action by combining resources. The ACCER provides the opportunity for Member agencies to leverage agency resources towards highest-priority issues of common concern, increase available expertise, increase support for high-priority management-oriented conservation research, and create economies of scale that would not otherwise be available to individual agencies acting alone; and

3. To ensure efforts to address issues of common concern are complementary and integrated.

The ACCER activities are supported by the Alaska Climate Change Coordinating Committee (C4) whose membership is comprised of ACCER member agency managers, science officers, or their equivalent, with sufficient authority to represent their agency on a policy level on a statewide basis. Responsibilities of the C4 group are to:

1. Integrate combined goals and science priorities of member agency climate change initiatives and efforts;

2. Provide interagency, management-level guidance to Member agency climate change efforts, including but not limited to Landscape Conservation Cooperatives (LCCs) and Climate Science Centers (CSC), and address cross-LCC considerations and decisions; and

3. Establish statewide goals and science priorities for CSC, and LCCs.

The Department of Interior's Alaska Climate Science Center

On February 22, 2010, Secretary of the Interior Ken Salazar issued an order that focused on coordinating the Department of Interior's (DOI) response to climate change. As a key component of this coordinated response strategy, the Secretarial order called for the creation of regional science centers to "provide climate change impact data and analysis geared to the needs of fish and wildlife managers as they develop adaptation strategies in response to climate change." These centers or "regional hubs" of coordinated climate change response were placed under the auspices of the National Climate Change and Wildlife Science Center (NCCWSC). In turn, NCCWSC was charged with the development of these centers in close cooperation with Interior agencies and other Federal, State, university, and non-governmental partners.

In early 2011, the USGS Alaska Climate Science Center (Alaska CSC) opened as the first of eight regional Climate Science Centers. The Alaska CSC is hosted by the University of Alaska Fairbanks (UAF) and is physically housed within the University of Alaska Anchorage (UAA). In addition to its relationship with UAF and UAA, the Alaska CSC also has developed strong partnerships with faculty from the University of Alaska Southeast, the U.S. Fish and Wildlife Service, National Oceanic and Atmospheric Administration,

U.S. Forest Service, and National Park Service. These partners provide expertise in climate science, ecology, impacts assessment, modeling, cultural impacts, and advanced information technology. These partnerships are essential for addressing climate issues in Alaska, where changes in temperature and precipitation are already having impacts on terrestrial and marine ecosystems.

As per the February 2010 Secretarial Order and the direction of its steering committee, the Alaska CSC is charged with providing scientific information, tools, and techniques that managers and other stakeholders interested in land, water, wildlife and cultural resources can use to anticipate, monitor, and adapt to climate change. Much of this work is aimed at meeting the needs of the Alaska Region Landscape Conservation Cooperatives, but most Alaska CSC activities also are intended to address the challenges faced by a much broader group of Federal, State, non-governmental organizations, and Alaskan Native entities.

Priority science activities of Alaska CSC include:

1. Use and creation of high-resolution climate models and derivative products to help forecast ecological change and population responses at local to regional scales;

2. Integration of physical climate models with ecological, habitat, and population response models; and

3. Development of methods to assess vulnerability of species, habitats, and human communities;

4. Development of standardized approaches to modeling, monitoring, data management and decision support.

The research direction taken by the Alaska CSC is guided by the Alaska Climate Change Executive Roundtable (ACCER), described above. The annual implementation of these science priorities is governed by the Alaska CSC Science Plan, which was completed in October 2011. Annual implementation of the science agenda is facilitated by the Alaska Climate Change Coordinating Committee (C4), a group that serves as a liaison to the various agency partners, while also reviewing the annual science plan and related project proposals.

Landscape Conservation Cooperatives

The Landscape Conservation Cooperatives (LCC) were formed in response to, and under the authority of, Section 3(c) of Secretarial Order 3289 (September 14, 2009): "Addressing the Impacts of Climate Change on America's Water, Land, and other Natural and Cultural Resources." Guidance provided to the LCCs included "...applied conservation science partnerships focused on a defined geographic area that inform on-the-ground strategic conservation efforts at landscape scales." Initiated by the U.S. Fish and Wildlife Service, LCCs, whose partners include DOI agencies, other Federal agencies, State agencies, Alaskan Native and Tribal

entities, non-governmental organizations, universities and others, are designed to foster collaboration and provide information and tools to land managers. In so doing, they seek to produce operational efficiencies and intellectual synergies in pursuit of science-based solutions to landscape-scale conservation. Within Alaska, LCCs are working as a network by cooperating on projects that span both LCC, State, Province, and international borders.

LCCs commonly are composed of partnerships dedicated to addressing landscape-level science and conservation information needs that commonly are beyond the ability of any single entity to adequately address. In Alaska, climate change as a landscape-level stressor takes center stage, but other potential climate related landscape-level stressors (for example, invasive species, contaminants, oil spills) also may be addressed. LCCs bring together scientists and managers, from diverse agencies and disciplines, to determine the landscape-level information that is most needed by land managers in their geographic area. The LCCs seek to deliver this information through collaborations, cooperative agreements, and coordination with institutions, such as the U.S. Geological Survey Climate Science Centers.

The five Landscape Conservation Cooperatives formed in Alaska (Aleutian and Bering Sea Islands, Arctic, North Pacific, Northwest Interior Forest, and Western Alaska) are part of a network of 22 LCCs covering all of the United States, most of Canada, and parts of Mexico.

Arctic LCC.—Due to polar amplification of climate change, this northernmost LCC is projected to experience more intense warming than any other and was among the first to be funded nationally. In its first two operational years, it has provided $2.6 million in funding (leveraging more than $5 million in partner contributions) to more than 30 research projects, addressing topics from the downstream effects of disappearing glaciers to climate-driven changes in invertebrate communities to the effects of a warming climate on food storage and prevalence of zoonotic diseases in subsistence foods. Models under development will allow for prediction of lake drainage risk and maternal den locations for polar bears, but also will integrate existing ecological response models to provide a more comprehensive view of what a future northern Alaska will look like under different climate scenarios. In the near term, the Arctic LCC will take a more systematic approach to addressing science needs by initiating linked inter-disciplinary studies focused on representative Arctic watersheds and eco-regional zones. The six technical working groups for this LCC (permafrost, geospatial, coastal process, species and habitat, hydrology, and climate modeling) provide technical input to the LCC's steering committee. This input is used in science planning and prioritization of research.

Western Alaska LCC.—The area represented by this LCC is underlain with intermittent permafrost and is largely bordered by low-relief coastline. As such, it is poised to undergo dramatic climate-driven ecological changes. The Western Alaska LCC was part of the second wave of LCC development, becoming operational in 2011. In its initial year of operation, it provided $1.3 million in funds (leveraging $1.8 million in partner contributions) to 12 projects that are intimately linked, and serve to inform each other as they progress, including five synthesized permafrost and hydrology projects. This LCC also is responding to information needs identified by the Western Arctic Caribou Herd Working Group by funding an investigation of the role of changing tundra on herd dynamics. Three coastal communities most affected by changing coastal processes will be the focus of climate change health assessments, courtesy of Western Alaska LCC funds. In 2012, this LCC is launching a Coastal Pilot Program, in which investigations will focus on coastal issues and science needs along Alaska's west coast.

North Pacific LCC.—The North Pacific LCC (NPLCC) includes the coastal temperate rainforest that extends along a narrow landscape corridor from the Kenai Peninsula in south-central Alaska, through British Columbia, to northwestern California. This rainforest generally is bound to the east by coastal mountain ranges and to the west by the Pacific Ocean. The NPLCC encompasses approximately 530,000 km^2 (204,000 mi^2) including more than 2,500 islands and is characterized by interconnected marine, freshwater, and terrestrial ecosystems, further linked by key species assemblages such as Pacific salmon and migratory birds. Strong human cultures, including numerous Tribes and First Nations, have thrived on the natural resources in this coastal margin since the last ice age, developing a rich body of Traditional Ecological Knowledge (TEK). Alaska segments of the coastal rainforest remain largely intact (90 percent) and are unique for their biodiversity, and the great ecological influences of glaciers in the region. Glacial runoff currently accounts for one-half the total runoff into the Gulf of Alaska and understanding the effects of climate change on hydrologic regimes and human economies are key goals of the NPLCC. A Strategic Science Plan is available and describes science support and new science priorities through 2017. In fiscal years 2011 and 2012, more than $1 million directly supported LCC data collection and information management, geospatial analysis, outreach and communication (website, workshops, and conferences), and projects designed to compile and synthesize existing information with special focus on TEK.

At the time of writing of this report, the Aleutian and Bering Sea Islands LCC and Northwestern Interior Forest LCC are in the early stages of development. Although neither has yet funded projects, both have functioning steering committees and are taking a strategic approach to science planning. Likewise, both are building capacity towards future collaborations in addressing pressing science and management needs throughout their geographic regions.

University of Alaska—Alaska Climate Research Center

The Alaska Climate Research Center was established and is funded by the State of Alaska under Title 14, Chapter 40, Section 085 (Alaska Climate Research Center, 2012). The primary mission of this center is to respond to inquiries concerning the meteorology and climatology of Alaska from public, private, and government agencies, and from researchers around the world. The center provides services within the three-tiered system (State, regional, and Federal). Most of the climatological data available for Alaska have been accumulated in Fairbanks and by the State climatologist in Anchorage at the Alaska State Climate Center. The Alaska Climate Research Center archives digital climate records, develops climate statistics, and writes monthly weather summaries, which are published in several newspapers around the State and in Weatherwise magazine. Research also is conducted on a number of high-latitude meteorological and climatological topics and provides useful links for related data.

Alaska Center for Climate Assessment and Policy

The Alaska Center for Climate Assessment and Policy (ACCAP; University of Alaska Fairbanks, 2012b) was established in 2006 with a mission to improve the ability of State and Federal agencies, industry, Tribal Governments and citizens in Alaska to respond to a changing climate. Funded by the National Oceanic and Atmospheric Administration (NOAA) Climate Program Office, ACCAP is one of several Regional Integrated Sciences and Assessments (RISA) programs nationwide. The RISA program supports research that addresses sensitive and complex climate issues of concern to decision makers and policy planners at a regional level. ACCAP fosters strong collaborations with university scientists, local experts in traditional knowledge, Federal, State and local planners, and members of industry and non-profit organizations. ACCAP is based at the University of Alaska Fairbanks and collaborates closely with the Scenarios Network for Alaska and Arctic Planning, the Alaska Climate Research Center and other groups to provide relevant and timely climate information statewide. Core program foci include coastal and living marine resources, community adaptation planning, Tribal impacts and adaptation, applied climate downscaling, wildfire, and sea ice.

The ACCAP model of "use-inspired science" engages scientists with end users in innovative research for informed decision making. Some of the decision support tools available from ACCAP include a quarterly climate newsletter, sea ice information tutorials, the development of a digital sea ice atlas, wildfire forecasts and other fire science tools, a growing suite of decision-support guidebooks, free monthly climate webinars, and digitally archived resources of Alaska-specific climate related topics.

State of Alaska Sub-Cabinet on Climate Change

In response to increasing impacts resulting from a warming climate, the Governor of Alaska issued Administrative Order 238, which established a Climate Change Sub-Cabinet. The Sub-Cabinet was tasked with developing a strategy to identify and ameliorate existing and projected climate effects. They also were charged with developing a list of research necessary to help better understand changes and inform decision makers. Perhaps the most critical work they undertook was addressing the urgent and near-term needs of communities most imperiled by flooding, erosion, and fires, for which they formed the Immediate Action Work Group.

The Climate Change Sub-Cabinet's efforts to draft a climate change strategy built on formative work, most notably that of the Alaska Climate Impact Assessment Commission, established by the Alaska Legislature in 2006. In an attempt to garner widespread input to the proposed strategy, an extensive process involving a broad spectrum of stakeholders was initiated in 2008. Participants comprised local, Federal and State agencies; Alaska Natives; and representatives of industry, academia, and environmental interest groups. The need for sustained coordination, within the State as well as with external partners, and leveraging of limited resources was an overarching theme that arose from the stakeholder process.

Stakeholders were divided into Adaptation and Mitigation Advisory Groups and corresponding technical work groups. Recommendations were made in the following areas concerning Adaptation: Public Infrastructure, Health and Culture, Natural Systems, the Opening of the Arctic, and Other Economic Activities. Measures that merited further consideration to reduce greenhouse gas emissions were made in sectors related to Energy Supply and Demand; Forestry, Agriculture, and Waste; Oil and Gas; Transportation and Land Use; and a Cross-Cutting set of issues. Specific recommendations can be found in reports from the Adaptation Advisory Group, the Mitigation Advisory Group, the Immediate Action Work Group, and the Research Needs Work Group (State of Alaska, 2011).

From late 2007 through early 2011, the Immediate Action Working Group made significant progress in providing assistance to six high-priority communities by providing a novel interagency approach. They engaged locals in overcoming numerous institutional barriers to mobilize State and Federal funds to build protective structures to defend against the forces of coastal erosion and provided assistance to some communities with efforts related to relocation or migration, as well as emergency and evacuation plans and training.

The State continues to take action today through numerous initiatives, some led by the State, such as the Northern Waters Task Force, formed by the legislature to gather information from Alaskans and other experts for the purpose of informing decision makers on important issues in the Arctic Ocean and other waters of the north, and others by the Federal Government. The State of Alaska has taken on key roles in a variety of interagency efforts, which include the Landscape Conservation Cooperatives and the Alaska Climate Change Executive Roundtable.

Alaska Native Tribal Health Consortium—Center for Climate and Heath

The Center for Climate and Health (Alaska Native Tribal Health Consortium, 2012a) was established in 2008 to assist the Alaska Tribal Health System in understanding the impacts of climate change and to help develop healthy adaptation practices. Special focus areas include damage and disruption to water and sanitation infrastructure, and evaluating vulnerability for water and food insecurity.

The Center is part of the Alaska Native Tribal Health Consortium (2012a), the statewide tribal health organization for Alaska, which is responsible for managing health facilities, providing clinical care and community health services, and for the design and construction of health infrastructure. It is staffed by individuals representing clinical health, community health and environmental health and engineering professions. The Center provides services that evaluate, monitor and help plan adaptation to climate change impacts. This includes assessments, surveillance, planning and funding improvements to operations and infrastructure. Education and outreach also are central to the Center's services as Alaskans seek to understand the human dimension of a rapidly changing environment. Some of these services include:

1. Health impact assessments;

2. Public health bulletins;

3. Adaptation plans;

4. Local Environmental Observer program;

5. Rural Alaska Monitoring Program for subsistence foods;

6. Engineering environmental atlas;

7. Climate and Health E-News;

8. Technical assistance, trainings and workshops;

9. Community demonstration project grants; and

10. Monthly Circumpolar Climate Change Events (Incidents) Maps.

Since 2009, the Center has performed a series of climate change health impact assessments in Northwest Alaska communities, and in 2012, the Center will expand these programs to Southwestern Alaska and the North Slope. In each community, impacts have been identified that affect public health, and completion of local assessment is the first step in developing effective plans and partnerships. Funding for the Center has been provided by the U.S. Environmental Protection Agency, the Indian Health Service, the Centers for Disease Control, and the U.S. Fish and Wildlife Service.

Local Environmental Observer Program

The Local Environmental Observer program (LEO; Alaska Native Tribal Health Consortium, 2012a) is composed of local Tribal environmental professionals who participate in a statewide climate change impact surveillance network. The program is open to local and regional Tribal professionals, who are environmental, environmental health and natural resource experts. The LEOs combine local and traditional knowledge with modern technology to record and document a wide range of local impacts that are climate-related or potentially climate-relevant.

The program is managed by the Center for Climate and Health at the Alaska Native Tribal Health Consortium. The LEOs are employed at local Tribal governments or for regional Tribal health organizations. The LEOs participate in statewide conferences and trainings for continuing education, and network in their communities with other local knowledge keepers and experts. LEOs provide time- and location-specific observations about local changes, on a variety of categories including:[7]

Extreme weather	Ice and cryosphere change
Air quality	Insects
Water security	Birds
Food security	Land animals
Seasons	Marine animals
Land change	Fish
Lake change	Shellfish
Coastal change	Infrastructure

The maps are updated monthly and are archived at the Center for Climate and Health for impact and trend analysis. The purpose of the program is to raise awareness about the types of climate-change related impacts that are occurring in Alaska communities, and to develop the dialogue between

local leadership and other stakeholders including government, academia, funding agencies and climate experts. Additionally, the LEO program acts as an observation system facilitator, referring LEOs to technical experts, outside resources and other monitoring and observation programs so as to further engage on specific issues that concerns their communities.

Center for Ocean Acidification

The University of Alaska-Fairbanks created an Ocean Acidification Research Center (University of Alaska Fairbanks, 2012g) within the School of Fisheries and Ocean Sciences. The mission of the Center is to:

1. Conduct research into ocean acidification, particularly in Alaskan waters, and determine the broader climate forcing's that are leading to decreases in ocean pH and the impacts of these changes on commercial species. The research will focus on three areas:

    a. Long-term autonomous monitoring and modeling efforts;

    b. Field observations in highly sensitive areas; and

    c. Quantifying physiological responses of vulnerable and commercially valuable species.

2. Maintain a central repository for the Federal and State Governments, as well as the public and private sectors, to access information relevant to ocean acidification and its impacts on fisheries and other economic resources.

Scenarios Network for Alaska and Arctic Planning

Scenarios Network for Alaska and Arctic Planning (SNAP; University of Alaska Fairbanks, 2012a) is a collaborative network of the University of Alaska, State, Federal, and local agencies, and non-governmental organizations. Using a select group of global models that perform best in northern latitudes, as well as Parameter-elevation Regressions on Independent Slopes Model climatology (Oregon State University, 2012), SNAP provides downscaled projections of future climatic conditions throughout Alaska. The primary products of the network are (1) datasets and maps projecting future conditions for selected variables, such as temperature, precipitation and growing season length, (2) rules and models that develop these projections, based on historical conditions and trends, and (3) explanations of uncertainty associated with these projections.

---

[7]Observations and photographs are posted on a web-accessible Google© Map, at www.anthc.org/chs/ces/climate/leo/ assessed on August 10, 2012 .

Currently (2012), most policy and management planning for Alaska and elsewhere assumes that future conditions will be similar to those of our recent past, however, there is reasonable consensus within the scientific community that future climatic, ecological, and economic conditions will likely be quite different from those of the past. We now know enough about current and likely future trajectories of climate and other variables to develop credible projections. We also can make projections for other variables that are closely correlated, such as frequency of intense storms, risk of wildfire or flooding, and habitat and wildlife changes associated with these events.

Alaska Ocean Observing System

The Alaska Ocean Observing System (AOOS; 2012a) is the regional ocean observing system for Alaska, providing observations, data and information products to meet agency and stakeholder needs. AOOS was formed in 2003 under a Memorandum of Agreement that establishes a governing board of Federal and State agencies and research institutions in Alaska. AOOS focuses on developing information products in four thematic areas, all of which are influenced by climate change: marine operations; coastal hazards; ecosystems, water quality and fisheries; and climate variability and trends.

A key AOOS product is its ocean and coastal data portal (Alaska Ocean Observing System, 2012b), which integrates real-time observations, model forecasts and remote sensing, and project data into information products for use by stakeholders. Specific information products related to climate change include: development of an historical sea ice atlas with information dating back to 1850, support for ocean acidification monitoring, support for long-term series of oceanographic measurements in the Chukchi Sea and in the Gulf of Alaska; and development of a "State of Alaska's Coasts and Oceans" electronic report.

AOOS is part of national and global networks of ocean observing, as codified by the Integrated Coastal and Ocean Observing System Act of 2009.

The goals of AOOS are to:

1. Support national defense, marine commerce, navigation safety, weather, climate and marine forecasting, energy siting and production, economic development, ecosystem-based management, public safety and public outreach, training and education;

2. Promote greater public awareness and stewardship of the Nation's ocean and coastal resources and the general public welfare;

3. Enable advances in scientific understanding to support the sustainable use, conservation, management and understanding of healthy ocean and coastal resources; and

4. Improve the Nation's capability to measure, track, explain and predict events related directly and indirectly to weather and climate change, natural climate variability, and interactions between the oceanic and atmospheric environments.

To achieve these goals, AOOS will:

1. Identify priorities for coastal and ocean observations and information based on the needs of users of Alaska's coasts and oceans;

2. Coordinate State, Federal, local, and private interests at a regional level to meet the priority needs of user groups in the Alaska region;

3. Identify gaps in existing ocean observing activities and data, make recommendations for needed increases to both Federal and non-Federal assets, and fill gaps when appropriate;

4. Increase efficiencies of existing ocean observing activities and data;

5. Increase the usefulness of ocean observations for a wider variety of users; and

6. Integrate observations and data through data management, planning, coordination and facilitation.

North Slope Science Initiative

The North Slope Science Initiative (NSSI; 2012) was authorized in Section 348 of the Energy policy Act of 2005 (Public Law 109-58). It is funded by Federal, State, and local governments with trust responsibilities for land and ocean management to facilitate and improve collection and dissemination of ecosystem information pertaining to the Alaskan North Slope region, including coastal and offshore regions. This information will be used to improve scientific and regulatory understanding of terrestrial, aquatic and marine ecosystems for consideration in the context of resource development activities and climate change. The NSSI strategic framework provides resource managers with the data and analyses they need to help evaluate multiple simultaneous goals and objectives related to each agency's mission on the North Slope of Alaska. The NSSI uses and complements both internal and external information produced under other North Slope science programs. The NSSI also facilitates information sharing among agencies, non-governmental organizations, industry, academia, international programs and members of the public to increase communication and reduce redundancy among science programs.

The NSSI has two advisory groups: the Oversight Group and the Science Technical Advisory Panel.

1. Oversight Group:

Activities of the NSSI are directed by the Oversight Group, which is composed of Federal, State, and Borough land managers. The Oversight Group consists of the following member agencies with voting privileges: Bureau of Land Management; Fish and Wildlife Service, National Park Service, National Marine Fisheries Service, Minerals Management Service, Alaska Department of Natural Resources, Alaska Department of Fish and Game, Arctic Slope Regional Corporation, and North Slope Borough. Additionally, the U.S. Geological Survey, Department of Energy, U.S. Arctic Research Commission, and National Weather Service participate in the Oversight Group as advisory agencies, but do not have voting privileges.

2. Science Technical Advisory Panel:

The purpose of the Science Technical Advisory Panel is to advise the NSSI Oversight Group on science issues such as identifying and prioritizing inventory, monitoring and research needs, and providing other scientific information as requested by the NSSI Oversight Group.

Planning for the Future

For this report, the U.S. Global Change Research Program (USGCRP) requested each of the Regions to identify a "path forward" for assessment activities that could be initiated and to provide a more sustained process in gathering, assessing, and reporting information into the next report. This also would include an assessment of the resource requirements needed to sustain the process in the future. The Report Teams also were asked to identify key research and data priorities for filling gaps in ongoing programs and providing support for adaptation and mitigation decisions. For the Alaska Regional report, a number of different potential activities could be initiated, assessed, and synthesized, some with specific goals or outcomes. These are condensed into three major areas of involvement or actions (fig. 45) and are summarized below (table 8).

Networking and Future Strategizing

In order to sustain an on-going climate assessment process, focused and concerted efforts should be directed toward maintaining a regional network of climate change players, communicating with USGCRP, and continuing to strategically respond to the rapidly evolving landscape of climate change science and services in Alaska. Regional players include

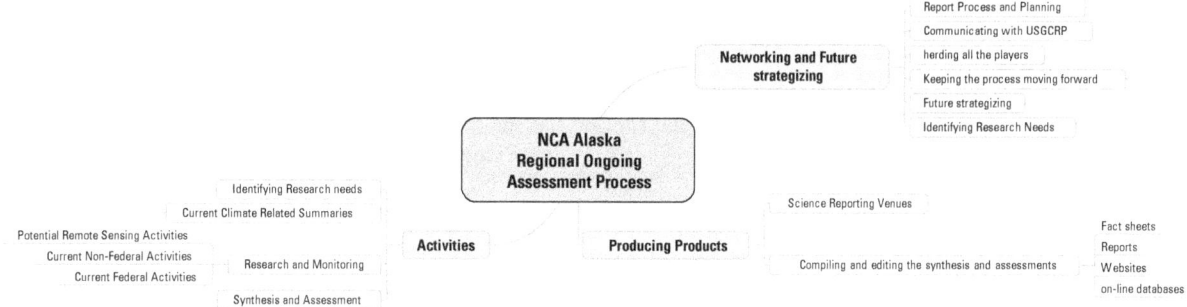

Figure 45.   Areas of ongoing activities that would provide authoritative information for use in future National Climate Assessment regional and national reports. NCA, National Climate Assessment; USGCRP, U.S. Global Change Research Program.

Table 8.   Annual venues that could be used to collect and disperse climate related information useful towards the Alaska Technical Regional report.

General venues	Specific venues
Alaska Forum on the Environment	Alaska Marine Science Symposium
Alaska Tribal Conference on Environmental Management	Alaska Federation of Natives
Bureau of Indian Affairs Providers Conference	Alaska Coastal Rainforest Symposium
Arctic Observation Network	U.S. Environmental Protection Agency Alaska Tribal Newsletter
	The Kachemak Bay Science Conference
	Western Alaska Science Conference

Federal, State, university, and non-profit scientists who are conducting on-going scientific research and monitoring. They also include Federal, State, and Tribal decisionmakers who require information for decisionmaking as well as the organizations and entities that have been formed to respond to changing environment, such as those mentioned above. It also includes people and entities who have taken on climate change adaptation and planning. Dedicated staff would need to interface between regional teams and the USGCRP. An on-going process would require continual evaluation not only of the existing climate change science, but also of the process as a whole.

To help ensure an ongoing process of collecting information and allow for public input, a number of different venues are available throughout the year (table 8). Some of these are more general in nature (for example, the Alaska Forum on the Environment), while others focus on specific groups (for example, Alaska Marine Science Symposium, Alaska Federation of Natives [representing Tribes] and Bureau of Indian Affairs Providers Conference). Special sessions, invited speakers, and information booths could be present at these venues to provide outreach opportunities to inform of, and request input to, both the regional report and the National report. Many of these venues, including special webinars directly related to the National Climate Assessment process, also provide opportunities to collect new information about the changing Alaskan environment as well as to make contact with potential contributing authors when special subject matter experts are desired.

Ongoing Activities

Ongoing and potential monitoring activities were identified that could be used as sustained monitoring or informing activities for reporting trends during the 4-year interval between reports. Existing monitoring activities that were recognized as having special importance include continuing measurement of glacier mass balance and documentation of ocean acidification. New activities that could be developed at selected sites include: continuous snow condition monitoring (distribution, onset and melt, density and depth); lake monitoring (area and depth); changes in stream and river temperature and chemistry [organic and inorganic compounds and potential toxins (for example, mercury)], especially over smaller watersheds where more dramatic changes in permafrost is taking place; distributed permafrost monitoring (temperature and depth profiles); monitoring of coastal erosion rates at high incident locations; and monitoring of coastal currents (direction, speed, and chemistry).

Another important area is the inclusion of traditional ecological knowledge (TEK) in both assessments of changes in the landscape (terrestrial and marine, including sea ice), and incorporation of that knowledge into traditional western science methodologies. Some of this knowledge has been recorded for some areas of interest [for example, University of Alaska Project Jukebox (University of Alaska Fairbanks, 2012c)], and also has been used to confirm physical events

such as dramatic lake drainage on the Alaskan Arctic Coastal Plain (Eisner and others, 2009). The use of new methodologies, however, such as Bayesian modeling, could provide new ways to integrate TEK with standard observations of scientific measurements as a means to produce new scenarios and models of ecological and biophysical changes across the Alaskan landscape and ocean environments. Respect for local knowledge and communication with Tribal groups is an important part of a complete assessment in Alaska.

In addition, it would be useful to conduct an assessment, of the on-going monitoring of reported data and information needs identified by State and Federal agencies and Tribal entities. This could include how the various land- and resource-management agencies access and obtain data, how they identify information needs and how they make decisions based on the available data. An identified gap to be filled here is compiling and sharing existing information identified and needs across agencies.

A number of direct data sources as well as existing literature reviews either summarize current climatic conditions or report on special events that may be directly or indirectly impacted by a changing climate. For example,

1. Alaska Center for Climate Assessment and Policy produces the Alaska Climate Dispatch (University of Alaska Fairbanks, 2012d) and Alaska Weather and Climate Highlights (University of Alaska Fairbanks, 2012e) that provide a synopsis of sub-regional and seasonal climate conditions and events;

2. The Alaska Native Tribal Health Consortium produces a weekly email of 'Climate and Health E-News' that highlights recent climate related events both locally and globally (Alaska Native Tribal Health Consortium, 2012b), and the Local Environmental Observer Network (mentioned previously) consisting of local environmental professionals from more than 60 communities across Alaska, that post local climate observations to a shared Google© Map available on the internet;

3. The National Oceanic and Atmospheric Administration (NOAA) produces two sources of useful information:

   a. The Arctic Report Card, which provides environmental information on the current state of the Arctic relative to historical records (National Oceanic and Atmospheric Administration, 2011);

   b. The Climate Prediction Center, which delivers climate prediction, monitoring, and diagnostic products for timescales from weeks to years (National Oceanic and Atmospheric Administration, 2012c);

4. The National Snow and Ice Data Center produces an Arctic Sea Ice News and Analysis and Monthly Highlights on recent events (National Snow and Ice Data Center, 2012);

5. The North Pacific Marine Science Organization (2012) produces a regular "State of the North Pacific" volume that summarizes trends of ocean variables over a 5-year period;

6. NOAA Fisheries Science Center produces for the North Pacific Fishery Management Council an annual status report on ocean conditions and endangered and threatened species referred to as the SAFE document, which is used in ecosystem-based management of commercial fisheries.

Information services such as these could provide a rich database of local and global trends that involve both a human and bio-physical dimension. These data sources could be summarized on a quarterly or yearly basis and used as input to the various sections of the report.

Within the Department of Interior–the National Park Service (2012a), U.S. Fish and Wildlife Service and the Bureau of Land Management (University of Alaska Anchorage, Alaska Natural Heritage Program 2012b) all have on-going or new inventory, monitoring and/or ecological assessment programs, from which information could be summarized (for example, over the last 4 or more years) to assess trends that have occurred across the Alaskan landscape. Some of these trends might involve key indicator species (for example, National Park Service Inventory and Monitoring Program); trends in migratory species (arrival, nesting success, etc.; U.S. Fish and Wildlife Service Migrating bird surveys) or changes in landscape condition (for example, Bureau of Land Management Rapid Landscape Assessment). The U.S. Geological Survey also conducts real-time permafrost and climate monitoring in Arctic Alaska (U.S. Geological Survey, 2012a); research on understanding landscape change in the recent (last 50 years) and distant (last 20,000 years) past (U.S. Geological Survey, 2012b); research to understand and project changes in marine and terrestrial ecosystems of the Arctic (Geiselman and others, 2012); and is developing new tools to map subsurface permafrost (Abraham, 2011). The National Park Service has undertaken a series of six Climate Change Scenario Planning workshops across Alaska in conjunction with Global Business Systems and Scenarios Network for Alaska and Arctic Planning, and with participation by multiple Federal, State, Tribal, local, and nongovernmental organizations. These science-based workshops are designed to anticipate potential future effects of climate change to the National Parks, park-affiliated communities and surrounding areas, identify areas of relative certainty and uncertainty, and identify a wide range of appropriate adaptation steps (National Park Service, 2012b).

Within the Department of Agriculture, the U.S. Forest Service has several projects examining climate change in the coastal temperate rainforest. The projects represent a range of development of climate change vulnerability assessments and assessment of mitigation options in ecosystems of the North Pacific coast. Forested ecosystem assessments include a vulnerability assessment of yellow-cedar that has experienced a widespread decline in the coastal temperate rainforest. This assessment also provides a framework for evaluating the influence of snow and soil moisture across forested landscapes to evaluate potential impacts on other plant species. Forest carbon studies have been conducted on unharvested and harvested stands to provide estimates of carbon flux rates across landscapes. The carbon assessments include carbon accretion rates in aboveground biomass and soil flux rates. The carbon studies incorporate both comprehensive plot measurements and extensive forest inventory analyses. Hydrologic studies include both evaluations of discharge and stream dependent organisms. Forest Service research is examining how stream temperatures and flow impact the timing of fry emergence and interactions with other organisms in stream systems. An extensive stream hydrograph model is available for determining rates of change in discharge under future climate scenarios. Forest Service research also is collaborating on glacial change studies with the University of Alaska to understand glacial mass change and associated hydrography, biogeochemistry, and interactions with marine systems.

In addition to the Federal activities in Alaska, a number of State of Alaska, University, and non-governmental entities also are conducting projects whose findings could be of use to future climate assessment reports. Some of these include:

1. State of Alaska has a Climate Change Sub-Cabinet that advises the Office of the Governor on the preparation and implementation of an Alaska climate change strategy (State of Alaska Administrative Order 238; State of Alaska, 2011);

2. Cook Inlet Keeper (2012) has developed the Stream Temperature Monitoring Network to build the science-based knowledge needed to identify thermal impacts in Alaska's coastal salmon habitat by collecting consistent, comparable temperature data for Cook Inlet's salmon streams. This network could be used as a long-term monitoring program that would provide valuable information to changes in salmon habitat;

3. Alaska Ocean Observing System (2012a) is developing an Ocean and Coastal Portal that integrates real-time data streams, remote sensing and model forecasts, and geo-spatially referenced data layers that include current and historical data. One major element underway is an electronic Sea Ice Atlas that will include historical sea ice data since the 1850s;

4. University of Alaska Fairbanks Permafrost Laboratory (University of Alaska Fairbanks, 2012f) deals with scientific questions related to the circumpolar permafrost dynamics and feedbacks between permafrost and global change. Data related to the thermal and structural state of circumpolar permafrost is collected and analyzed;

5. National Phenology Network (2012) brings together citizen scientists, government agencies, non-profit groups, educators, and students of all ages to monitor the impacts

of climate change on plants and animals in the United States. Although in its infant stage in Alaska, this network could provide unique insights to changes throughout the State;

6. Oil and gas companies including Shell (2012), ConocoPhillips (2012), BP (2012), and Statoil (2012) are conducting numerous research and monitoring projects on the North Slope and in the Chukchi and Beaufort Seas. Public access to this data is increasing;

7. Alaska Coastal Rainforest Center (2012) has an annual science symposium, is developing an Alaska Citizens Science Network, a collaborative effort between the Alaska Coastal Rainforest Center and University of Alaska Southeast to provide resources and educational applications to citizen science projects throughout Alaska, and conducts periodic BioBlitz's to survey different watersheds each year to provide a better understanding of citizens observations of local biodiversity;

8. The Alaska Native Tribal Health Consortium, Center for Climate and Health (2012a) conducts community-scale climate change health impact assessments, and provides technical assistance and monitoring for climate-related health effects and food and water security.

Because of Alaska's great geographic extent and varied ecosystems, the use of remote sensing may prove to be a promising means to both monitor and detect changes across the landscape, especially in current and past time frames being observed (that is, from the present back to the 1970s). In addition, it may provide interdisciplinary science opportunities a more synoptic view of the interactions between earth and climate systems, natural communities and ecosystems, and human infrastructure. Existing and potential remote sensing products are currently (2012) available over a number of spatial scales. For example, at coarse resolutions (250–1,000 m) enhanced Moderate Resolution Imaging Spectro-radiometer (eMODIS) satellite data (Jenkerson and others, 2010; U.S. Geological Survey, 2012c) can be used for creating a community-specific suite of vegetation and land cover monitoring products. These may include (at varying resolutions) vegetation indices (such as Normalized Difference Vegetation Index); leaf area index and gross primary productivity. Some land-cover characteristics include fire, land cover type and dynamics, and ecosystem performance.

The current U.S. Landsat database (U.S. Geological Survey, 2012d) at a medium resolution (30–50 m) also contains a wealth of satellite imagery for Alaska that could be used for long-term (1970s–present) analysis of changes to the landscape, such as coastal erosion (Mars and Houseknecht, 2007; Jones and others, 2008) and land cover mapping (Shasby and Carneggie, 1986; Talbot and Markon, 1986, 1988; Markon, 1992, 1995; Markon and Wesser, 1996; Selkowitz and Stehman, 2011). Other remote sensing systems include the SPOT satellite (Satellite Imaging Corporation, 2012), Quickbird (Digital Globe, 2012), and Ikonos (GeoEye Satellites, 2012) and various airborne Lidar and Radar systems

over specific sites, but are usually used for a one-time data-collection effort. A new source of remote sensing data is the High Frequency Radars currently (2012) mapping in real-time sea surface currents in the Chukchi Sea (U.S. Bureau of Ocean Energy Management, 2012). These are funded by the Bureau of Ocean Energy Management, industry, and the Alaska Ocean Observing System, and can be used for operational activities, such as oil spill response and search and rescue, as well as providing historical climatological data. All of these remote sensing systems, considered either collectively or individually, have potential to provide a record of physical (for example, coastal erosion, winds, ocean color, sea surface temperature, and precipitation) or biological (for example, land cover) process rates and change across the time dimension.

An on-going assessment process also will necessarily entail regularly compiling and synthesizing all of the following: research needs, current climate related summaries, and Federal and non-Federal research and monitoring. Once compiled and synthesized, this information will need to be assessed, or critically evaluated by a credible group of experts with broad disciplinary experience in a balanced and transparent way to guide policy decisions. This process requires familiarity with regionally specific scientific research as well as an understanding of relevant regional stakeholders and existing climate science networks. The synthesis and assessment part of the process pulls all of the relevant information together in a way that can be useful to policy makers and as such is a vital step in an on-going process. It also will require dedicated staff and resources.

Communication Resources

An on-going synthesis process, such as the one envisioned by U.S. Global Change Research Program (USGCRP), will require not only synthesis and assessment of information and resources, but the production of printed, on-line, and other communication media that convey assessment results to important audiences, such as Congress, State and Federal agencies, and other regional stakeholders. This may include on-line data repositories and data bases as well as web portals for "one-stop-shopping" in climate science and information. Production of these communication products will require skilled and knowledgeable science writers to produce fact sheets, on-line web content, and potential multi-media, such as videos that accurately convey assessment results in a language and context that is readily applied in public policy and resource management. It also will require graphics design and technical programming skills and resources for both data management and web-site design and production.

The process by which the next regional technical report is organized and initiated should be accomplished shortly after the report is submitted to the USGCRP. During the initial meetings, the previous year's writing team should consider accomplishing the following actions:

1. Review activities, events, and processes used during the prior report writing process that provides a lessons learned summary;

Ecosystem Performance

Using satellite data, methods have been developed for analyzing ecosystems performance over large geographic landscapes. One such method utilizes the normalized difference vegetation index (or NDVI), which is a measure of greenness that can be interpreted in terms of plant growth or photosynthetic activity. By calculating the growing season integral of NDVI over multiple years, the status and trends of changes across ecosystems can be calculated. When analyzed with other known factors such as fire and land management practices, this information can be used to separate the influences of climate from other types of change processes. In the figure below, an ecosystem performance anomaly for 2004 is shown, with perimeters of fires that occurred between 1997 and 2004 (Wylie and others, 2008).

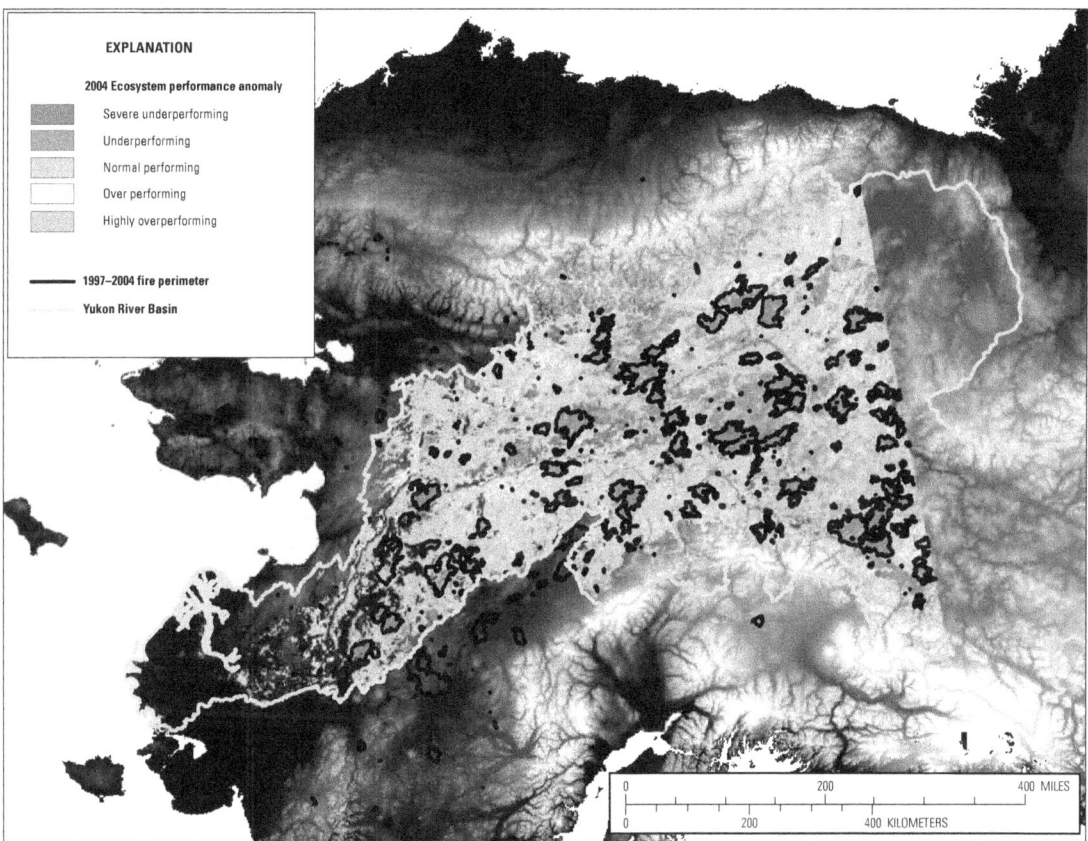

Ecosystem performance anomaly for 2004, with perimeters of fires that occurred between 1997 and 2004. The colored portion of the image represents the boreal forest areas of Alaska within the Yukon River Basin.

2. Identify and provide initial organization to the next writing team and select one or more team leads;

3. Establish a draft timeline of events that provides for an ongoing identification and analysis of potential technical information as it becomes available and ensures public input from a variety of venues;

4. Develop a proposed budget and identify potential sources of funding; and

5. Identify data access, utilization, indexing, retrieval, and archiving means and processes.

It is important that this activity be accomplished in collaboration with a broad range of partners to ensure that a wide range of experience from different sectors is represented to provide input to the next report.

References Cited

Aagaard, K., and Carmack, E.C., 1989, The role of sea ice and other fresh water in the arctic circulation: Journal of Geophysical Research, v. 94, no. C10, p. 14,485–14,498.

Aagaard, K., Coachman, L.K., and Carmack, E.C., 1981, On the halocline of the Arctic Ocean: Deep Sea Research Part A, v. 28, no. 6, p. 529–545.

Aagaard, K., Weingartner, T.J., Danielson, S.L., Woodgate, R.A., Johnson, G.C., and Whitledge, T.E., 2006, Some controls on flow and salinity in Bering Strait: Geophysical Research Letters, v. 33, L19602, doi:10.1029/2006GL026612.

Abdul-Aziz, O.I., Mantua, N.J., and Myers, K.W., 2011, Potential climate change impacts on thermal habitats of Pacific salmon (*Oncorhynus* spp.) in the North Pacific Ocean and adjacent seas: Canadian Journal of Fisheries and Aquatic Science, v. 65, no. 9, p. 1,660–1,680.

Abraham, J., 2011, A promising tool for subsurface permafrost mapping—An application of airborne geophysics from the Yukon River Basin, Alaska: U.S. Geological Survey, Fact Sheet 2011-3133, 4 p.in marine and terrestrial ecosystems of the Arctic: U.S. Geological Survey Fact Sheet 2011-3136, 4 p. (Also available at http://pubs.usgs.gov/fs/2011/3133/.)

Alaska Climate Impact Assessment Commission, 2008, Final Commission report: Alaska State Legislature, March 17, 2008, accessed June 7, 2012, at http://www.housemajority.org/coms/cli/cli_finalreport_20080301.pdf.

Alaska Climate Research Center, 2009, Temperature change in Alaska: Alaska Climate Research Center website, posted May 23, 2012, accessed June 7, at http://climate.gi.alaska.edu/ClimTrends/Change/TempChange.html

Alaska Climate Research Center, 2012, website, accessed August 10, 2012, at http://climate.gi.alaska.edu/.

Alaska Coastal Rainforest Center, 2012, website, accessed August 10, 2012, at http://acrc.alaska.edu/.

Alaska Department of Environmental Conservation, 2010a, Imperiled community water resource analysis: Prepared by Tetra Tech, Anchorage, Alaska, p. v–vi, accessed July 23, 2012, at http://www.climatechange.alaska.gov/docs/iaw_tt_imperiled_h2o_30jun10.pdf.

Alaska Department of Environmental Conservation, 2010b, Health and culture, chap. 7 of Alaska's climate change strategy, addressing impacts in Alaska, accessed June 7, 2012, at http://www.climatechange.alaska.gov/aag/aag.htm.

Alaska Department of Environmental Conservation, 2012, Climate change in Alaska: Alaska website, accessed June 8, 2012, at http://www.climatechange.alaska.gov/.

Alaska Department of Natural Resources, Division of Forestry, 2010, What's bugging Alaska's forests? Spruce bark beetle facts and figures: website, accessed August 10, 2012, at http://forestry.alaska.gov/insects/sprucebarkbeetle.htm.

Alaska Miners Association, Inc., 2008, The economic impacts of Alaska's mining industry: Alaska Miners Association, Inc., accessed June 8, 2012, at http://www.alaskaminers.org/mcd07sum.pdf.

Alaska Native Tribal Health Consortium, 2012a, Center for Climate and Health, website, accessed August 10, 2012, at http://www.anthc.org/chs/ces/climate/.

Alaska Native Tribal Health Consortium, 2012b, Climate and Health E-News, website, accessed August 10, 2012, at http://www.anthc.org/chs/ces/climate/listserv.cfm.

Alaska Native Tribal Health Consortium, 2012c, Local Environmental Observer (LEO) Network, websites, accessed September 17, 2012, at www.anthc.org/chs/ces/climate/leo.

Alaska Ocean Observing System, 2012a, website, accessed August 10, 2012, at www.aoos.org.

Alaska Ocean Observing System, 2012b, Data Portal, website accessed August 10, 2012, at http://data.aoos.org/maps/sensors/#1=sensor-stations.

Alaska Regional Assessment Group, 1999, The potential consequences of climate variability and change–Alaska: Prepared for the U.S. Global Change Research Program, Center for Global Change and Arctic System Research, University of Alaska Fairbanks, Fairbanks, Alaska, accessed June 8, 2012, at http://www.globalchange.gov/what-we-do/assessment/previous-assessments/the-first-national-assessment-2000/605.

Alekseev, G., Danilov, A., Kattsov, V., Kuz'mina, S., and Ivanov, N., 2009, Changes in the climate and sea ice of the Northern Hemisphere in the 20th and 21st centuries from data of observations and modeling: Izvestiya Atmospheric and Oceanic Physics, v. 45, no. 6, p. 675–686.

Alessa, L., Kliskey, A., Lammers, R., Arp, C., White, D., Hinzman, L., and Busey, R., 2008, The arctic water resource vulnerability index—an integrated assessment tool for community resilience and vulnerability with respect to freshwater: Environmental Management, v. 42, no. 3, p. 523–541.

Amstrup, S.C., DeWeaver, E.T., Douglas, D.C., Marcot, B.G., Durner, G.M., Bitz, C.M., and Bailey, D.A., 2010, Greenhouse gas mitigation can reduce sea-ice loss and increase polar bear persistence: Nature, v.468, p. 955–960.

Amstrup, S.C., Marcot, B.G., and Douglas, D.C., 2008, A Bayesian network modeling approach to forecasting the 21st century worldwide status of polar bears, *in* DeWeaver, E.T., Bitz, C.M., and Tremblay, L.B., eds., Arctic sea ice decline—Observations, projections, mechanisms, and implications: Geophysical Monograph Series 180, American Geophysical Union, Washington D.C., p. 213–268.

Arctic Climate Impact Assessment, 2004, Impacts of a warming Arctic—Arctic Climate Impact Assessment: Cambridge University Press, 146 p., accessed June 8, 2012, at http://www.acia.uaf.edu/pages/overview.html.

Arctic Climate Impact Assessment, 2005, Arctic Climate Impact Assessment: Cambridge University Press, 1,042 p., accessed June 8, 2012, at http://www.acia.uaf.edu/pages/scientific.html.

Arctic Council, 2009, Arctic Marine Shipping Assessment 2009 Report: Tromsø, Norway, Protection of the Arctic Marine Environment (PAME) Working Group, 194 p., accessed June 8, 2012, at http://www.pame.is/amsa/amsa-2009-report.

Arctic-Yukon-Kuskokwim Sustainable Salmon Initiative, 2012, website, accessed September 17, 2012, at http://www.aykssi.org/.

Arendt, A., Echelmeyer, K.A., Harrison, W.D., Lingle, C.S., and Valentine, V., 2002, Rapid wastage of Alaska glaciers and their contribution to rising sea level: Science, v. 297, no. 5580, p. 382–386, doi:10.1126/science.1072497.

Arendt, A., Walsh, J., and Harrison, W., 2009, Changes of glaciers and climate in northwestern North America during the late twentieth century: Journal of Climate, v. 22, no. 15, p. 4,117–4,134.

Arrigo, K. R., van Dijken, G., and Pabi, S., 2008, Impact of a shrinking arctic ice cover on marine primary production: Geophysical Research Letters, v. 35, L19603, doi:10.1029/2008GL035028.

Arrigo, K.R., and van Dijken, G., 2011, Secular trends in Arctic Ocean net primary production: Journal of Geophysical Research, v. 116, C09011, 15 p., doi:10.1029/2011JC007151.

Arsan, E.L., 2006, Potential for dispersal of the non-native parasite *myxobolus cerebralis*–Qualitative risk assessments for the State of Alaska and the Willamette River basin, Oregon: Corvallis, Oregon State University, Master's Thesis, accessed June 8, 2012, at http://scholarsarchive.library.oregonstate.edu/xmlui/bitstream/handle/1957/3843/Arsan.pdf?sequence=7.

AVISO, 2012, Mean Sea Level Rise, website, accessed August 10, 2012, at http://www.aviso.oceanobs.com/en/news/ocean-indicators/mean-sea-level/.

Azetsu-Scott, K., Clarke, A., Falkner, K., Hamilton, J., Jones, E.P., Lee, C., Petrie, B., Prinsenberg, S., Starr, M., and, Yeats, P., 2010, Calcium carbonate saturation states in the waters of the Canadian Arctic Archipelago and the Labrador Sea: Journal of Geophysical Research—Oceans, v. 115, C11021, 18 p., doi:10.1029/2009JC005917.

Azumaya, T., Nagasawa, T., Temnykh, O.S., and Khen, G.V., 2007, Regional and seasonal differences in temperature and salinity limitations of Pacific salmon: North Pacific Anadromous Fish Commission, v. 4, p. 179–187.

Bailey, R.G., 1983, Delineation of ecosystem regions: Environmental Management, v. 7, no. 4, p. 365–373

Bailey, R.G., 1985, The factor of scale in ecosystem mapping: Environmental Management, v. 9, no. 4, p. 271–276.

Bailey, R.G., 1988, Ecogeographic analysis: A guide to the ecological division of land for resource management: U.S. Department of Agriculture, U.S. Forest Service, Miscellaneous Publication 1465, 18 p.

Bailey, R.G., Avers, P.E., King, T., and McNab, W.H., 1994, Ecoregions and subregions of the United States: U.S. Department of Agriculture, U.S. Forest Service, map, scale 1:7,500,000.

Balshi, M.S., McGuire, A.D., Duffy, P., Flannigan, M., Walsh, J., and Melisso, J., 2009, Assessing the response of area burned to changing climate in western boreal North America using a multivariate adaptive regression splines (MARS) approach: Global Change Biology, v. 15, no. 3, p. 578–600.

Bamber, J., and Riva, R., 2010, The sea level fingerprint of recent ice mass fluxes: The Cryosphere, v. 4, p. 621–627, doi:10.5194/tc-4-621-2010.

Barber, V.A., Juday, G.P., and Finney, B.P., 2000, Reduced growth of Alaskan white spruce in the twentieth century from temperature–induced drought stress: Nature, v. 405, no. 6787, p. 668–673.

Barnes, P.W., Rearic, D.M., and Reimnitz, E., 1984, Ice gouging characteristics and processes *in* Barnes, P.W., Schell, D.M., and Reimnitz, E., eds., The Alaskan Beaufort Sea—Ecosystems and Environment, New York, Academic Press, p. 185–212.

Barrett, K., McGuire, A.D., Hoy, E.E., and Kasischke, E.S., 2011, Potential shifts in dominant forest cover in interior Alaska driven by variations in fire severity: Ecological Applications, v. 21, no. 7, p. 2,380–2,396.

Bates, N.R., and Mathis, J.T., 2009, The Arctic Ocean marine carbon cycle—Evaluation of air-sea CO_2 exchanges, ocean acidification impacts, and potential feedbacks: Biogeosciences, v. 6, p. 2,433–2,459.

Bates, N.R., Mathis, J.T., and Cooper, L.W., 2009, Ocean acidification and biologically induced seasonality of carbonate mineral saturation states in the western Arctic Ocean: Journal of Geophysical Research, v. 114, C11007, 21 p., doi:10.1029/2008JC004862.

Beamish, R.J., and Bouillon, D.R., 1993, Pacific salmon production trends in relation to climate: Canadian Journal of Fisheries and Aquatic Science, v. 50, p. 1,002–1,016.

Beamish, R.J., Noakes, D.J., McFarlane, G.A., Klysahtorin, L., Ivanov, V.V., and Kurashov, V., 1999, The regime concept and natural trends in the population of Pacific salmon: Canadian Journal of Fisheries and Aquatic Science, v. 56, no. 3, p. 516–526.

Beier, C.M., Lovecraft, A.L., and Chapin, F.S., III, 2009, Growth and collapse of a resource system—An adaptive cycle of change in public lands governance and forest management in Alaska: Ecology and Society, v. 14, no. 2, 5 p., accessed June 8, 2012, at http://www.ecologyandsociety.org/vol14/iss2/art5/.

Beier, C.M., Sink, S.E., Hennon, P.E., D'Amore, D.V., and Juday, G.P., 2008, Twentieth-century warming and the dendroclimatology of declining yellow-cedar forests in southeastern Alaska: Canadian Journal of Forest Research, v. 38, no. 6, p. 1,319–1,334.

Beltaos, S., and Burrell, G.C., 2003, Climatic change and river ice breakup: Canadian Journal of Civil Engineering, v. 30, no. 1, p. 145–155.

Bengtsson, L., Hodges, K.I., and Roeckner, E., 2006, Storm tracks and climate change: Journal of Climate, v. 19, no. 15, p. 3,518–3,543.

Berg, E.E., Hillman, K.M., Dial, R., and DeRuwe, A., 2009, Recent woody invasion of wetlands on the Kenai Peninsula Lowlands, south–central Alaska—a major regime shift after 18,000 years of wet Sphagnum–sedge peat recruitment: Canadian Journal of Forest Research, v. 39, no. 11, p. 2,033–2,046.

Berner, J., and Furgal, C., 2005, Human health, chap. 15 of Arctic Climate Impact Assessment: Cambridge, United Kingdom, and New York, Cambridge University Press, p. 863–906, accessed June 8, 2011, at http://www.acia.uaf.edu/pages/scientific.html.

Bernhardt, E.L., Hollingsworth, T.N., and Chapin, F.S., III, 2011, Fire severity mediates climate-driven shifts in understorey community composition of black spruce stands of interior Alaska: Journal of Vegetation Science, v. 22, no. 1, p. 32–44, doi:10.1111/j.1654-1103.2010.01231.x.

Berns, H., 2010, A framework for the preliminary assessment of vulnerability in fisheries–dependent communities: Corvallis, Oregon State University, Department of Sociology, master's thesis, 75 p.

Berthier, E., Schiefer, E., Clarke, G.K.C., Menounos, B., and Rémy, F., 2010, Contribution of Alaskan glaciers to sea-level rise derived from satellite imagery: Nature Geoscience, v. 3, p. 92–95, doi:10.1038/ngeo737.

Bhatia, M., Das, S., Longnecker, K., Charette, M., and Kujawinski, E., 2010, Molecular characterization of dissolved organic matter associated with the Greenland ice sheet: Geochimica Et Cosmochimica Acta, v.74, p.3768–3784.

Bieniek, P., Bhatt, U., Rundquist, L., Lindsey, S., Zhang, X., and Thoman, R., 2010, Large-scale climate controls of interior Alaska river ice breakup: Journal of Climate, v. 24, p. 286–297.

Bigler, B.S., Welch, D.W., and Helle, J. H., 1996, A review of size trends among North Pacific salmon (Oncorhynchus spp.): Canadian Journal of Fisheries and Aquatic Science, v. 53, p.455–465.

Bowden, W.B., Gooseff, M.N., Balser, A., Green, A., Peterson, B.J., and Bradford, J.H., 2008, Sediment and nutrient delivery from thermokarst features in the foothills of the North Slope, Alaska—Potential impacts on headwater stream ecosystems: Journal of Geophysical Research, v. 113, G02026, 12 p., doi:10.1029/2007JG000470.

Boyd, P.W., Law, C.S., Wong, C.S., Nojiri, Y., Tsuda, A., Levasseur, M., Takeda, S., Rivkin, R., Harrison, P.J., Strzepek, R., Gower, J., McKay, R.M., Abraham, E., Arychuk, M., Barwell-Clarke, J., Crawford, W., Crawford, D., Hale, M., Harada, K., Johnson, K., Kiyosawa, H., Kudo, I., Marchetti, A., Miller, W., Needoba, J., Nishioka, J., Ogawa, H., Page, J., Robert, M., Saito, H., Sastri, A., Sherry, N., Soutar, T., Sutherland, N., Taira, Y., Whitney, F., Wong, S.K.E., and Yoshimura, T., 2004, The decline and fate of an iron-induced subarctic phytoplankton bloom: Nature, v. 428, p. 549–553.

BP, 2012, Alaska, website, accessed August 10, 2012, at http://www.bp.com/sectiongenericarticle.do?categoryId=9030181&contentId=7055693.

Brabets, T.P., and Walvoord, M.A., 2009, Trends in streamflow in the Yukon River basin from 1944 to 2005 and the influence of the Pacific Decadal Oscillation: Journal of Hydrology, v. 371, p. 108–119.

Brinkman, T.J., Hansen, W., BurnSilver, S., Kofinas, G., and Chapin, T., 2011a, Subsistence resource availability project, Venetie, Alaska: University of Alaska–Fairbanks, Scenarios Network for Alaska Planning.

Brinkman, T.J., Hansen, W., Chapin, T., and Kofinas, G., 2011b, Subsistence resource availability project, Fort Yukon, Alaska: University of Alaska–Fairbanks, Scenarios Network for Alaska Planning.

Bromirski, P.D., Miller, A.J., Flick, R.E., and Auad, G., 2011, Dynamical suppression of sea level rise along the Pacific coast of North America—Indications for imminent acceleration: Journal of Geophysical Research, v. 116, C07005, doi:10.1029/2010JC006759.

Bronen, R., 2011, Climate-induced relocations–Creating an adaptive governance framework based in human rights doctrine: New York University Review of Law and Social Change, v. 35, p, 356–406.

Brown, R., Derksen, C., and Wang, L., 2010, A multi-data set analysis of variability and change in Arctic spring snow cover extent, 1972–2008: Journal of Geophysical Research, v. 115, D1611, 16 p., doi:10.1029/2010JD013975.

Brown, R.D., and Mote, P.W., 2009, The response of northern hemisphere snow cover to a changing climate: Journal of Climate, v. 22, p. 2,124–2,145.

Brubaker, M., Bell, J., Berner, J., and Warren, J., 2011, Climate change health assessment, a novel approach for Alaska Native communities: International Journal of Circumpolar Health, v. 70, no. 3, p. 266–273.

Brubaker, M., Bell, J., and Rolin, A., 2009b, Climate change effects on traditional Inupiat food cellars: Anchorage, Alaska, Alaska Native Tribal Health Consortium, 7 p., accessed November 16, 2011, at http://www.anthc.org/chs/ces/climate/upload/CCH-Bulletin-No-01-Permafrost-and-Underground-Food-Cellars-Revised-Final.pdf.

Brubaker, M., Berner, J., Bell, J., and Warren, J., 2009a, Climate change in Point Hope, Alaska, strategies for community health: Alaska Native Tribal Health Consortium, Center for Climate and Health, accessed July 23, 2012, at http://www.anthc.org/chs/ces/climate/upload/Climate-Change-and-Health-Effects-in-Point-Hope-Alaska.pdf.

Brubaker, M., Berner, J., Bell, J., and Warren, J., 2010, Climate change in Kivalina, Alaska, strategies for community health: Alaska Native Tribal Health Consortium, Center for Climate and Health, accessed July 23, 2012, at http://www.anthc.org/chs/ces/climate/upload/Climate-Change-in-Kivalina-Alaska-Strategies-for-Community-Health-2.pdf.

Bryant, M.D., 2009, Global climate change and potential effects on Pacific salmonids in freshwater ecosystems of southeast Alaska: Climatic Change, v. 95, no. 1-2, p.169–193, doi:10.1007/s10584-008-9530-x.

Caldeira, K., and Wickett, M.E., 2005, Ocean model predictions of chemistry changes from carbon dioxide emissions to the atmosphere and ocean: Journal of Geophysical Research–Oceans, v. 110, C09S04, 12 p., doi:10.1029/2004JC002671.

Calef, M.P., McGuire, A.D., and Chapin, F.S., III, 2008, Human influences on wildfire in Alaska from 1988 through 2005—An analysis of the spatial patterns of human impacts: Earth Interactions, v. 12, no. 1, p. 1–17, doi10.1175/2007EI220.1.

Callaway, D., Eamer, J., Edwardsen, E., Jack, C., Marcy, S., Olrun, A., Patkotak, M., Rexford, D. and Whiting, A., 1999, Effects of climate change on subsistence communities in Alaska, in Weller, G., and Anderson, P.A., eds., Assessing the consequences of climate change for Alaska and the Bering Sea region: University of Alaska–Fairbanks, Center for Global Change and Arctic Systems Research, p. 59–73.

Carlson, M.L., and Lapina, I., 2004, Invasive non–native plants in the Arctic—The intersection between natural and anthropogenic disturbance: Poster presentation at the American Academy of Science meeting, Anchorage, Alaska.

Carlson, M.L., and Shephard, M., 2007, Is the spread of non-native plants in Alaska accelerating?, in Meeting the challenge—Invasive plants in Pacific Northwest ecosystems: Portland, Oreg., U.S. Forest Service, Pacific Northwest Research Station, En. Tech. Rep. PNW-GTR-694, p. 111–127, accessed June 8, 2011, at http://www.uaa.alaska.edu/enri/publications/upload/GTR-694_MLC_MS.pdf.

Carmack, E., and Chapman, D., 2003, Wind-driven shelf/basin exchange on an Arctic shelf—The joint roles of ice cover extent and shelf-break bathymetry: Geophysical Research Letters, v. 30, 4 p., doi:10.1029/2003GL017526.

Carothers, C., Lew, D.K., and Sepez, J., 2010, Fishing rights and small communities—Alaska halibut IFQ transfer patterns: Ocean and Coastal Management, v. 53, no. 9, p. 518–523.

Cassano, J.J., Uotila, P., and Lynch, A.H., 2006, Changes in synoptic weather patterns in the polar regions in the 20th and 21st centuries—part 1, Arctic: International Journal of Climatology, v. 26, no. 8, p.1,027–1,049, doi: 10.1002/joc.1306.

Cavalieri, D.J., Parkinson, C.L., and Vinnikov, K.Y., 2003, 30-year satellite record reveals contrasting Arctic and Antarctic decadal sea ice variability: Geophysical Research Letters, v. 30, no. 1970, 4 p., doi:10.1029/2003GL018031.

Cazenave, A., and Llovel, W., 2010, Contemporary sea level rise: Annual Review of Marine Science, v. 2, p. 145–173.

Cerveny, L.K., 2004, Preliminary research findings from a study of the social-cultural effects of tourism in Haines, Alaska: General Technical Report PNW-GTR-612, U.S. Department of Agriculture, Forest Service, Pacific Northwest Research Station, Portland, Oreg., 144 p., accessed June 8, 2012, at http://www.fs.fed.us/pnw/pubs/pnw_gtr612.pdf.

Chapin, F.S., III, McGuire, A.D., Ruess, R.W., Hollingsworth, T.N., Mack, M.C., Johnstone, J.F., Kasischke, E.S., Euskirchen, E.S., Jones, J.B., Jorgenson, M.T., Kielland, K., Kofinas, G.P., Turetsky, M.R., Yarie, J., Lloyd, A.H., and Taylor, D.L., 2010, Resilience of Alaska's boreal forest to climatic change: Canadian Journal of Forest Research, v. 40, p. 1,360–1,370.

Chapin, F.S., III, Oswood, M.W., Van Cleve, K., Viereck, L.A., and Verbyla, D.L., 2006, Alaska's Changing Boreal Forest: Oxford, United Kingdom, Oxford University Press.

Chapin, F.S., III, Trainor, S.F., Huntington, O., Lovecraft, A.L., Zavaleta, E., Natcher, D.C., McGuire, A.D., Nelson, J.L., Ray, L., Calef, M., Fresco, N., Huntington, H.P., Rupp, T.S., DeWilde, L., and Naylor, R.L., 2008, Increasing wildfire in the boreal forest—Causes, consequences, and pathways to the potential solutions of a wicked problem: BioScience, v. 58, no. 6, p. 531–540.

Cherry, J., Walker, S., Fresco, N., Trainor, S., and Tidwell, A., 2010, Impacts of climate change and variability on hydropower in southeast Alaska—Planning for a robust energy future: International Arctic Research Center website, accessed June 8, 2012, at http://www.iarc.uaf.edu/research/highlights/2010/impacts-on-hydropower-southeast.

Chierici, M., and Fransson, A., 2009, Calcium carbonate saturation in the surface water of the Arctic Ocean—undersaturation in freshwater influenced shelves: Biogeosciences, v. 6, no. 11, p. 2,421–2,432.

Chinowsky, P., Strzepek, K., Larsen, P., and Opdahl, A., 2010, Adaptive climate response cost models for infrastructure: Journal of Infrastructure Systems, v. 16, no. 3, p. 173–180, doi:10.1061/(ASCE)IS.1943-555X.0000021.

City-Data, 2012, Alaska—Agriculture: Website, accessed June 8, 2012, at http://www.city-data.com/states/Alaska-Agriculture.html.

Cleland, D.T., Avers, P.E., McNab, W.H., Jensen, M.E., Bailey, R.G., King, T., and Russell, W.E., 1997, National hierarchical framework of ecological units, in Boyce, M.A., and Haney, A., eds., Ecosystem management applications for sustainable forest and wildlife resources: New Haven, Conn., Yale University Press, p. 181–200.

Clow, G.D., DeGange, A.R., Derksen, D.V., and Zimmerman, C.E., 2011, Climate change considerations, in Holland-Bartels, Leslie, and Pierce, Brenda, eds., 2011, An evaluation of the science needs to inform decisions on Outer Continental Shelf energy development in the Chukchi and Beaufort Seas, Alaska: U.S. Geological Survey Circular 1370, p. 81–108. (Also available at http://pubs.er.usgs.gov/publication/cir1370.)

Clow, G.D., and Urban, F.E., 2002, Large permafrost warming in northern Alaska during the1990s determined from GTN-P borehole temperature Measurements: Fall Meeting, Eos Transactions, American Geophysical Union, v. 83, no. 47, Fall Meeting Supplement, Abstract B11E-04, December 6–10, 2002, San Francisco.

Coachman, L.K., Aagaard, K., and Tripp, R.B., 1975, Bering Strait—The Regional Physical Oceanography: Seattle, University of Washington Press.

Code of Federal Regulations, Executive Order 13112, 1999, Invasive Species: Federal Register, v. 64, no. 25, p. 6,183–6,186, accessed June 8, 2012, at http://frwebgate.access.gpo.gov/cgi-bin/getdoc.cgi?dbname=1999_register&docid=99-3184-filed.pdf.

Cohen, S.C., and Freymueller, J.T., 2004, Crustal deformation in southcentral Alaska—The 1964 Prince William Sound Earthquake subduction zone: Advances in Geophysics, v. 47, p. 1–63.

Cohen, J.L., Furtado, J.C., Barlow, M.A., Alexeev, V.A., and Cherry, J.C., 2012, Arctic warming, increasing snow cover and widespread boreal winter cooling: Environmental Research Letters, v. 7, 014007, 8 p., doi:10.1088/1748-9326/7/1/014007.

Cole, H., Colonell, V., and Esch, D., 1999, The economic impact and consequences of global climate change on Alaska's infrastructure, in Assessing the consequences of climate change for Alaska and the Bering Sea region: University of Alaska–Fairbanks, Workshop proceedings summarized for the U.S. Global Change Research Program, p. 43–56, accessed June 8, 2012, at http://www.besis.uaf.edu/besis-oct98-report/Infrastructure-1.pdf.

Colt, S., Dugan, D., and Fay, G., 2007, The Regional economy of southeast Alaska: Prepared for the Alaska Conservation Foundation by the Institute of Social and Economic Research, University of Alaska–Anchorage, accessed June 8, 2012, at http://www.iser.uaa.alaska.edu/Publications/SoutheastEconomyOverviewfinal4.pdf.

Comiso, J., and Nishio, F., 2008, Trends in the sea ice cover using enhanced and compatible AMSR-E, SSM/I, and SMMR data: Journal of Geophysical Research, v. 113, C02S07, doi:10.1029/2007JC004257.

Commission for Environmental Cooperation, 1997, Ecological regions of North America—Toward a common perspective: Montreal, Commission for Environmental Cooperation, 71 p.

Conn, J.S., Beattie, K.L., Shepard, M.A., Carlson, M.L., Lapina, I., Hebert, M., Gronquist, R., Densmore, R., and Rasy, M., 2008, Alaska Melilotus invasions—Distribution, origin, and susceptibility of plant communities: Arctic, Antarctic, and Alpine Research, v. 40, no. 2, p. 298–308.

Connor, C., Steveler, G., Post, A., Monteith, D., and Howell, W., 2009, The neoglacial landscape and human history of Glacier Bay: The Holocene, v. 19, no. 3, p. 381–393, doi:10.1177/0959683608101389.

ConocoPhillips Alaska, 2012, website, accessed August 10, 2012, at http://alaska.conocophillips.com.

Conservation of Arctic Floara and Fauna, Circumploar Biodiversity Monitoring Program, 2012, website, accessed August 14, 2012, at http://www.caff.is/about-the-cbmp.

Cook Inlet Keeper, 2012, website, accessed August 10, 2012, at http://inletkeeper.org/.

Coronado, C., and Hilborn, R., 1998, Spatial and temporal factors affecting survival in coho salmon (Oncorhynchus kisutch) in the Pacific Northwest: Canadian Journal of Fisheries and Aquatic Science, v. 55, no. 9, p. 2,255–2,265.

Cortés-Burns, H., Lapina, I., Klein, S.C., Carlson, M.L., and Flagstad, L., 2008, Invasive plant species monitoring and control—Areas impacted by 2004 and 2005 fires in interior Alaska—A survey of Alaska BLM lands along the Dalton, Steese, and Taylor Highways: Report funded by the Bureau of Land Management, Alaska State Office, Anchorage, Alaska, 162 p., accessed July 20, 2012, at http://aknhp.uaa.alaska.edu/wp-content/uploads/2010/11/Cortes_etal_2008l.pdf.

Crusius, J., Schroth, A.W., Gasso, S., Moy, C.M., Levy, R.C., and Gatica, M., 2011, Glacial flour dust storms in the Gulf of Alaska—Hydrologic and meteorological controls and their importance as a source of bioavailable iron: Geophysical Research Letters, v. 38, L06602, 5 p., doi:10.1029/2010GL046573.

Curtis, J., Wendler, G., Stone, R., and Dutton, E., 1998, Precipitation decrease in the western Arctic, with special emphasis on Barrow and Barter Island, Alaska: International Journal of Climatology, v. 18, p. 1,687–1,707.

D'Amore, D.V., and Hennon, P.E., 2006, Evaluation of soil saturation, soil chemistry, and early spring soil and air temperatures as risk factors in yellow-cedar decline: Global Change Biology, v. 12, no. 3, p. 524–545.

Danielson, S.L., Curchitser, E.N., Hedstrom, K.S., Weingartner, T.J., and Stabeno, P.J., 2011, On ocean and sea ice modes of variability in the Bering Sea: Journal of Geophysical Research, v. 116, C12034, 24 p., doi:10.1029/2011JC007389.

de Rivera, C.E., Steves, B.P., Ruiz, G.M., Fofonoff, P., and Hines, A.H, 2007, Northward spread of marine nonindigenous species along western North America—Forecasting risk of colonization in Alaskan waters using environmental niche modeling: Report submitted to Prince William Sound Regional Citizens' Advisory Council and the U.S. Fish and Wildlife Service, 36 p., accessed June 8, 2012, at http://www.pwsrcac.org/docs/d0041100.pdf.

Delworth, T.L., Broccoli, A.J., Rosati, A., Stougger, R.J., Balaji, V., Beesley, J.A., Cooke, W.F., Dixon, K.W., Dunne, J., Dunne, K.A., Durachta, J.W., Findell, K.L., Ginoux, P., Gnanadesikan, A., Gordon, C.T., Griffies, S.M., Gudel, R., Harrison, M.J., Held, I.M., Hemler, R.S., Horowitx, L.W., Klein, S.A., Knutson, T.R., Malyshev, S.L., Milly, P.C.D., Ramaswamy, V., Russell, J., Schwarzkopf, M.D., Shevliankova, E., Sirutis, J.J., Spelman, M.J., Stern, W.F., Winton, M., Wittenberg, A.T., Wyman, B., Zeng, F., and Zhang, R., 2006, GFDL's CM2 global coupled climate models, part I—formulation and simulation characteristics: Journal of Climate, v. 19, p. 645–674.

Demain, J., Gessner, B., McLaughlin, J., Sikes, D., and Foote, J., 2009, Increasing insect reactions in Alaska—Is this related to climate change?: Allergy and Asthma Proceedings, v. 30, no .3, p. 238–243.

Denman, K.L., Mackas, D.L., Freeland, H.J., Austin, M.J., and Hill, S.H., 1981, Persistent upwelling and mesoscale zones of high productivity off the west coast of Vancouver Island, Canada, in: Richards, F. A., ed., Coastal Upwelling: Washington, D.C., American Geophysical Union.

Derksen, C., and Brown, R., 2011, Snow, in Arctic Report Card—Update for 2011: Environment Canada, Climate Research Division, accessed June 8, 2012, at http://www.arctic.noaa.gov/reportcard/snow.html.

Deser, C., and Teng, H., 2008, Evolution of Arctic sea ice concentration trends and the role of atmospheric circulation forcing, 1979–2007: Geophysical Research Letters, v. 35, L02504,doi:10.1029/2007GL032023.

Digital Globe, 2012, website, accessed August 10, 2012, at http://www.digitalglobe.com/.

Douglas, D.C., 2010, Arctic sea ice decline–Projected changes in timing and extent of sea ice in the Bering and Chukchi Seas: U.S. Geological Survey Open-File Report 2010–1176, 31 p. (Also available at http://pubs.er.usgs.gov/publication/ofr20101176.)

Downton, M.W., and Miller, K.A., 1998, Relationships between Alaskan salmon catch and North Pacific climate on interannual and interdecadal time scales: Canadian Journal of Fisheries and Aquatic Science, v. 55, no. 10, p. 2,067–2,077.

Duffy, P.A., Walsh, J.E., Mann, D.H., Graham, J.M., and Rupp, T.S., 2005, Impacts of large-scale atmospheric-ocean variability on Alaskan fire season severity: Ecological Applications, v. 15, no. 4, p. 1,317–1,330.

Duguay, C., Prowse, T., Bonsal, B., Brown, R., Lacroix, M., and Menard, P., 2006, Recent trends in Canadian lake ice cover: Hydrological Processes, v. 20, p. 781–801.

Dukes, J.S., and Mooney, H.A., 1999, Does global change increase the success of biological invaders?: Trends in Ecology and Evolution, v. 14, no. 4, p. 135–139.

Durner, G.M., Douglas, D.C., Nielson, R.M., Amstrup, S.C., McDonald, T.L., Stirling, I., Mauritzen, M., Born, E.W., Wiig, Ø., DeWeaver, E., Serreze, M.C., Belikov, S.E., Holland, M.M., Maslanik, J., Aars, J., Bailey, D.A., and Derocher, A.E., 2009, Predicting 21st-century polar bear habitat distribution from global climate models: Ecological Monographs, v. 79, p. 25–58.

Eisner, W.R., Cuomo, C.J., Hinkel, K.M., Jones, B.M., and Brower, R.H., 2009, Advancing landscape change research through the incorporation of Iñupiaq knowledge: Arctic, v. 62, no. 4, p. 429–442.

Elliott, J., Larsen, C.F., Freymueller, J.T., and Motyka, R.J., 2010, Tectonic block motion and glacial isostatic adjustment in southeast Alaska and adjacent Canada constrained by GPS measurements: Journal of Geophysical Research, v. 115, B09407, 21 p., doi:10.1029/2009JB007139.

Euskirchen, E.S., McGuire, A.D., and Chapin, F.S., III., 2007, Energy feedbacks of northern high-latitude ecosystems to the climate system due to reduced snow cover during 20th century warming: Global Change Biology, v. 13, no.11, p. 2,425–2,438, doi:10.1111/j.1365-2486.2007.01450.x.

Euskirchen, E.S., McGuire, A.D., Chapin, F.S., III, and Rupp, T.S., 2010, The changing effects of Alaska's boreal forests on the climate system: Canadian Journal of Forest Resources, v. 40, no. 7, p. 1,336–1,346.

Euskirchen, E.S., McGuire, A.D., Chapin, F.S., III, Yi, S., and Thompson, C.C., 2009, Changes in vegetation in northern Alaska under scenarios of climate change, 2003–2100—Implications for climate feedbacks: Ecological Applications, v. 19, no. 4, p. 1,022–1,043.

Fay, F.H., 1982, Ecology and biology of the Pacific walrus, *Odobenus rosmarus divergens Illiger*: North American Fauna, v. 74, p. 1-279.Fellman, J., Spencer, R.G.M., Edwards, R., D'Amore, D., Hernes, P.J., and Hood, E., 2010, The impact of glacier runoff on the biodegradability and biochemical composition of terrigenous dissolved organic matter in near-shore marine ecosystems: Marine Chemistry, v. 121, p. 112–122, doi:10.1016/j.marchem.2010.03.009.

Fellman, J., Spencer, R.G.M., Edwards, R., D'Amore, D., Hernes, P.J., and Hood, E., 2010, The impact of glacier runoff on the biodegradability and biochemical composition of terrigenous dissolved organic matter in near-shore marine ecosystems: Marine Chemistry, v.121, p.112-122, doi:10.1016/j.marchem.2010.03.009.

Fischbach, A.S., Amstrup, S.C., and Douglas, D.C., 2007, Landward and eastward shift of Alaskan polar bear denning associated with recent sea ice changes: Polar Biology, v. 30, p. 1,395–1,405.

Fischbach, A.S., Monson, D.H., and Jay, C.V., 2009, Enumeration of Pacific walrus carcasses on beaches of the Chukchi Sea in Alaska following a mortality event, September 2009: U.S. Geological Survey Open-File Report 2009–1291, 11 p. (Also available at http://pubs.er.usgs.gov/publication/ofr20091291.)

Flato, G.M., and Boer, G.J., 2001, Warming asymmetry in climate change simulations: Geophysical Research Letters, v. 28, no. 1, p. 195–198. doi:10.1029/2000GL012121.

Flato, G.M., and Hibler, W.D., III, 1992, Modeling pack ice as a cavitating fluid: Journal of Physical Oceanography, v. 22, p. 626–651.

Fleming, R.A., and Volney, W.J.A., 1995, Effects of climate change on insect defoliator population processes in Canada's boreal forest—Some plausible scenarios: Water, Soil, and Air Pollution, v. 82, p. 445–454.

Fluharty, D., 2005, Evolving ecosystem approaches to management of fisheries in the USA, *in* Browman, H.I., and Stergiou, K.I., eds., Politics and socio-economics of ecosystem-based approaches to the management of marine resources: Marine Ecology Progress Series, v. 275, p. 265–270.

Foster, M., and Goldsmith, S., 2008, Replacement cost for public infrastructure in Alaska—An update: Web Note No. 4: University of Alaska-Anchorage, Institute of Social and Economic Research, accessed June 8, 2012, at http://www.iser.uaa.alaska.edu/Publications/webnote/Web_Note4a.pdf.

Francis, O.P., 2011, Atmospheric forcing of wave states in the southeast Chukchi Sea: University of Alaska–Fairbanks, Ph.D. dissertation.

Francis, O.P., Panteleev, G.G., and Atkinson, D.E., 2011, Ocean wave conditions in the Chukchi Sea from satellite and in situ observation: Geophysical Research Letters, v. 38, L24610, 5 p., doi:10.1029/2011GL049839.

Francis, R.C., Hixon, M.A., Clarke, M.E., Murawski, S.A., and Ralston, S., 2007, Ten commandments for ecosystem-based fisheries scientists: Fisheries, v. 32, no. 5, p. 217–233.

Frei, A., and Gong, G., 2005, Decadal to century scale trends in North American snow extent in coupled atmosphere-ocean general circulation models: Geophysical Research Letters, v. 32, LI8502, 5 p., doi:10.1029/2005GL023394.

French, M.H., Goovaerts, F.,P., and Kasischke, E.S., 2004, Uncertainty in estimating carbon emissions from boreal forest fires: Journal of Geophysical Research, v. 109, D14S08, 12 p., doi:10.1029/2003JD003635.

Freymueller, J.T., Woodard, H., Cohen, S., Cross, R., Elliott, J., Larsen, C., Hreinsdottir, S., and Zweck, C., 2008, Active deformation processes in Alaska, based on 15 years of GPS measurements, in Freymueller, J.T., Haeussler, P.J., Wesson, R., and Ekstrom, G., eds., Active tectonics and seismic potential of Alaska: American Geophysical Union Geophysical Monograph 179, p. 1–42.

Fried, N., 2011, Alaska's $49 billion economy, in Schultz, C., and Fried, N., Gender and earnings in Alaska: Alaska Economic Trends, October 2011, v. 31, no. 10, p. 14–16, accessed July 20, 2012, at http://labor.alaska.gov/trends/trends2011.htm.

Friends of the Earth, 2012, Arctic Shipping, website, accessed August 10, 2012, at http://www.foe.org/projects/oceans-and-forests/oceangoing-vessels/arctic-shipping.

Gallant, A.L., Binnian, E.F., Omernik, J.M., and Shasby, M.B., 1995, Ecoregions of Alaska: U.S. Geological Survey Professional Paper 1567, 73 p. (Also available at http://pubs.er.usgs.gov/publication/pp1567.)

Gascard, J.C., Festy, J., Goff, H., Weber, M., Bruemmer, B., Offermann, M., Doble, M., Wadhams, P., Forsberg, R., Hanson, S., Skourup, H., Gerland, S., Nicolaus, M., Metaxian, J.-P., Grangeon, J., Jaapala, H., Rinne, E., Haas, C., Wegener, A., Heygster, G., Jakobson, E., Palo, T., Wilkinson, J., Kaleschke, L., Claffey, K., Elder, B., and Bottenheim, J., 2008, Exploring Arctic transpolar drift during dramatic sea ice retreat: EOS, v. 89, no. 3 p. 21, doi:10.1029/2008EO030001.

GeoEye Satellites, 2012, website, accessed August 10, 2012, at http://www.geoeye.com/CorpSite/products-and-services/imagery-sources/.

Geiselman, J., DeGange, T., Oakley, K., Derksen, D., and Whalen, M., 2012, Changing Arctic ecosystems—Research to understand and project changes in marine and terrestrial ecosystems of the Arctic: U.S. Geological Survey Fact Sheet 2011-3136, 4 p. (Also available at http://pubs.usgs.gov/fs/2011/3136/.)

Gessner B., 2008, Lack of piped water and sewage services is associated with pediatric lower respiratory tract infection in Alaska: Journal of Pediatrics, v. 152, no. 5, p. 666–670, DOI:10.1016/j.jpeds.2007.10.049.

Giles, K., Laxon, S., and Ridout, A., 2008, Circumpolar thinning of Arctic sea ice following the 2007 record ice extent minimum: Geophysical Research Letters, v. 35, L22502, 4 p., doi:10.1029/2008GL035710.

Gnanadesikan, A., Dixon, K.W., Griggies, S.M., Balaji, V., Barreiro, M., Beesley, J.A., Cooke, W.F., Delworth, T.L., Gerdes, R., Harrison, M.J., Held, I.M., Hurlin, W.J., Lee, H-C., Liang, Z., Nong, G., Pacanowski, R.C., Rosati, A., Russell, J., Samuels, B.L., Song, Q., Spelman, M.J., Stougger, R.J., Sweeney, C.O., Gabriel, V., Winton, M., Wittenberg, A.T., Zeng, F., Zhang, R., and Dunne, J.P., 2006, GFDL's CM2 global coupled climate models, part II–The baseline ocean simulation: Journal of Climate, v. 19, p. 675–697.

Goetz, S.J., Bunn, A.G., Fiske, G.J., and Houghton, R.A., 2005, Satellite-observed photosynthetic trends across boreal North America associated with climate and fire disturbance: Proceedings of the National Academy of Sciences, v. 102, p. 13,521–13,525.

Gordon, C., Cooper, C., Senior, C.A., Banks, H.T., Gregory, J.M., Johns, T.C., Mitchell, J.F.B. and Wood, R.A., 2000, The simulation of SST, sea ice extents and ocean heat transports in a version of the Hadley Centre coupled model without flux adjustments: Climate Dynamics, v.16, p.147-168.

Grebmeier, J.M., 2012, Shifting patters of life in the Pacific Arctic and sub–Arctic seas: Annual Review of Marine Science, v. 4, p. 63–78.

Grebmeier, J.M., Moore, S.E., Overland, J.E., Fray, K.E., and Gradinger, R., 2010, Biological response to recent Pacific Arctic sea ice retreats: Eos Transactions, American Geophysical Union, v. 91, no. 18, p. 161–163.

Grebmeier, J.M., Overland, J.E., Moore, S.E., Farley, E.V., Carmack, E.C., Cooper, L.W., Frey, K.E., Helle, J.H., McLaughlin, F.A., and McNutt, S.L., 2006, A major ecosystem shift in the northern Bering Sea: Science, v. 311, no. 5766, p. 1,461–1,464, doi:10.1126/science.1121365.

Grebmeier, J.M., Priscu, J.C., D'Arrigo, R., Ducklow, H.W., Fleener, C., Frey, K.E., and Rosa, C., 2011, Frontiers in understanding climate change and polar ecosystems—Summary of a workshop: Washington, D.C., National Academy Press, 78 p.

Griffith, B., and McGuire, A.D., 2008, National wildlife refuges case study—Alaska and the central flyway in Annex A of Julius, S.H., and West, J.M., eds., and Baron, J.S., Griffith, B., Joyce, L.A., Kareiva, P., Keller, B.D., Palmer, M.A., Peterson, C.H., and Scott, J.M, Preliminary review of adaptation options for climate-sensitive ecosystems and resources—A report by the U.S. Climate Change Science Program and the Subcommittee on Global Change Research: U.S. Environmental Protection Agency, p. A-36–A-46.

Grosse, G., Harden, J., Turetsky, M., McGuire, A.D., Camill, P., Tarnocai, C., Frolking, S., Schuur, E.A.G., Jorgenson, T., Marchenko, S., Romanovsky, V., Wickland, K.P., French, N., Waldrop, M., Bourgeau-Chavez, L., Striegl, R.G., 2011, Vulnerability of high-latitude soil carbon in North America to disturbance: Journal of Geophysical Research–Biogeosciences, v. 116, G00K06, 23 p., doi:10.1029/2010JG001507.

Hannah, C.G., Dupont, F., and Dunphy M., 2009, Polynyas and tidal currents in the Canadian Arctic Archipelago, Arctic, v. 62 no. 1, p. 83–95.

Hanson, E., and Sytsma, M.D., 2007, The potential for mitten crab *Eriocheir sinensis* H Milne Edwards, 1853 (Crustacea: Brachyura) invasion of Pacific Northwest and Alaskan estuaries: Biological Invasions, v. 10, no. 5, p. 603–614, doi:*10.1007/s10530-007-9156-3.*

Hare, S.R., and Mantua, N.J., 2000, Empirical evidence for North Pacific regime shifts in 1977 and 1989: Progress in Oceanography, v. 47, no. 2–4, p. 103–145.

Hare, S.R., Mantua, N.J., and Francis, R.C., 1999, Inverse production regimes - Alaska and West Coast Pacific Salmon: Fisheries, v.24, p.6-14.

Hasumi, H., and Emori, S., eds., 2004, K-1 Coupled GCM (MIROC) description—K-1 Technical Report No. 1: Tokyo, Japan, University of Tokyo Center for Climate System Research, National Institute for Environmental Studies, and Frontier Research Center for Global Change, accessed June 12, 2012, at http://www.ccsr.u-tokyo.ac.jp/kyosei/hasumi/MIROC/tech-repo.pdf.

Heintz, R.A., and Vollenweider, J.J., 2010, Influence of size on the sources of energy consumed by overwintering walleye pollock (*Theragra chalcogramma*): Journal of Experimental Marine Biology and Ecology, v. 393, p. 43–50.

Hennessy, T., Ritter, T., Holman, R., Bruden, D., Yorita, K., Bulkow, L., Cheek, J., Singleton, R., and Smith, J., 2008, The relationship between in-home water service and the risk of respiratory tract, skin, and gastrointestinal tract infections among rural Alaska natives: American Journal of Public Health, v. 98, p. 2,072–2,078.

Hennon, P.E., D'Amore, D., Wittwer, D., Johnson, A., Schaberg, P., Hawley, G., Beier, C., Sink, S., and Juday, G., 2006, Climate warming, reduced snow, and freezing injury could explain the demise of yellow-cedar in southeast Alaska, USA: World Resource Review, v. 18, no. 2, p. 427–450.

Hennon, P.E., Shaw, C.G., and Hansen, E.M., 1990, Dating decline and mortality of *Chamaecyparis nootkatensis* in Southeast Alaska: Forest Science, v. 36, no. 3, p. 502–515.

Herman-Mercer, N., Schuster, P.F., and Maracle, K.B., 2011, Indigenous observations of climate change in the lower Yukon river basin, Alaska: Human Organization, v. 70, no. 3, p. 244–252.

Hierro, J.L., Villareal, D., Eren, O., Graham, J.M., and Callaway, R.M., 2006, Disturbance facilitates invasion—The effects are stronger abroad than at home: American Naturalist, v. 168, no. 2, p. 144–156.

Hinch, S.G., Cooke, S.J., Healey, M.C., and Farrell, A.P., 2005, Behavioural physiology of fish migrations—salmon as a model approach, *in* Sloman, K.A., Wilson, R.W., and Balshine, S.,. eds., Behaviour and physiology of fish: Fish Physiology Series, v. 24, p. 240-285.

Hines, A.T., Ruiz, G.M., Hitchcock, N.G., and DeRivera, C., 2004, Projecting range expansion of invasive European green crabs (*Carcinus maenas*) to Alaska—Temperature and salinity tolerance of larvae: Final Report to Prince William Sound Regional Citizens' Advisory Council, accessed June 8, 2011, at http://www.pwsrcac.org/docs/d0018700.pdf.

Hinzman L.D., Bettez, N.D., Bolton, W.R., Chapin, F.S., Dyurgerov, M.B., Fastie, C.L., Griffith, B., Hollister, R.D., Hope, A., Huntington, H.P., Jensen, A.M., Jia, G.J., Jorgenson, T., Kane, D.L., Klein, D.R, Kofinas, G., Lynch, A.H., Lloyd, A.H., McGuire, A.D., Nelson, F.E., Oechel, W.C., Osterkamp, T.E., Racine, C.H., Romanovsky, V.E., Stone, R.S., Stow, D.A., Sturm, M., Tweedie, C.E., Vourlitis, G.L., Walker, M.D., Walker, D.A., Webber, P.J., Welker, J.M., Winker, K.S., and Yoshikawa, K., 2005, Evidence and implications of recent climate change in northern Alaska and other arctic regions: Climatic Change, v. 72, p. 251–298.

Hollowed, A.B., A'Mar, T., Barbeaux, S., Bond, N., Ianelli, J.N., Spencer, P., and Wilderbuer, T., 2011, Integrating ecosystem aspects and climate change forecasting into stock assessments: Alaska Fisheries Science Center Quarterly Research Report (July, August, September), National Oceanic and Atmospheric Administration, National Marine Fisheries Service, 6 p., accessed June 8, 2012, at http://www.afsc.noaa.gov/Quarterly/jas2011/JAS11feature.pdf

Hood, E., and Berner, L., 2009, Effects of changing glacial coverage on the physical and biogeochemical properties of coastal streams in southeastern Alaska: Journal of Geophysical Research, v. 114, G03001, doi:10.1029/2009JG000971.

Hood, E., Fellman, J., Spencer, R.G.M., Hernes, P.J., Edwards, R., D'Amore, D., and Scott, D., 2009, Glaciers as a source of ancient and labile organic matter to the marine environment: Nature, v. 462, p. 1,044–1,048, doi:10.1038/nature08580.

Hood, E., and Scott, D., 2008, Riverine organic matter and nutrients in southeast Alaska affected by glacial coverage: Nature Geoscience, v. 1, p. 583–587, doi:10.1038/Ngeo280.

Hopkins, D.M., and Hartz, R.W., 1978, Coastal morphology, coastal erosion, and barrier islands of the Beaufort Sea, Alaska: U.S. Geological Survey Open-File Report 78–1063, 54 p. (Also available at http://pubs.er.usgs.gov/publication/ofr781063.)

Hu, F.S., Barnes, J., and Rupp, S., 2010a, Reconstructing fire regimes in tundra ecosystems to inform a management-oriented ecosystem model: Final Report for Joint Fire Science Program Project Number 06-3-1-23, CESU Agreement J979106K153/001, April 2010, 26 p., accessed July 20, 2012, at http://www.firescience.gov/projects/06-3-1-23/project/06-3-1-23_hu_et_al_finalreport_jfsp_06-3-1-23.pdf.

Hu, F.S., Higuera, P.E., Walsh, J.E., Chapman, W.L., Duffy, P.A., Brubaker, L.B., and Chipman, M.L., 2010b, Tundra burning in Alaska—Linkages to climatic change and sea ice retreat: Journal of Geophysical Resources, v. 115, G04002, 8 p., doi:10.1029/2009JG001270.

Hunt, J., Rosenberger, A., Daum, D., and Solie, D., 2008, Changes in water temperature from three locales along the Yukon River: 2008 American Association for the Advancement of Science Arctic Science Conference, accessed June 8, 2012, at http://arcticaaas.org/meetings/2008/2008_aaas_abstracts_v1.4.pdf.

Hunt, G.L., Jr., Stabeno, P., Walters, G., Sinclair, E., Brodeur, R.D., Napp, J.M., and Bond, N.A., 2002, Climate change and control of the southeastern Bering Sea pelagic ecosystem: Deep-Sea Research, part II, v. 49, no. 26, p. 5,821–5,854.

Hunter, C.H., Caswell, H., Runge, M.C., Regehr, E.V., Amstrup, S.C., and Stirling, I., 2010, Climate change threatens polar bear populations—A stochastic demographic analysis: Ecology, v. 91, p. 2,883–2,897.

Huntington, H.P., Kruse, S.H., and Scholz, A.J., 2009, Demographics and environmental conditions are uncoupled in the Pribilof Islands social ecological system: Polar Research, v. 28, no. 1, p. 119–128.

Intergovernmental Panel on Climate Change, 2007, Climate Change 2007—Synthesis report—Contribution of Working Groups I, II, and III to the Fourth Assessment Report of the Intergovernmental Panel on Climate Change: Geneva, Switzerland, 104 p., accessed June 8, 2012, at http://www.ipcc.ch/publications_and_data/publications_ipcc_fourth_assessment_report_synthesis_report.htm.

Ishida, Y., Ito, S., Kaeriyama, M., McKinnell, S., and Nagasawa, K., 1993, Recent changes in age and size of chum salmon (Oncorhynchus keta) in the North Pacific Ocean and possible causes: Canadian Journal of Fisheries and Aquatic Science, v. 50, no. 2, p. 290–295.

Ishida, Y., Ito, S., and Murai, K., 1995, Density dependent growth of pink salmon (Oncorhynchus gorbuscha) in the Bering Sea and western North Pacific: North Pacific Anadromous Fish Commission Doc. 140, National Research Institute of Far Seas Fisheries, 5-7-1 Orido, Shimizu, Shizuoka, 424 Japan, 17 p.

Jackson, S.T., Betancourt, J.L., Booth, R.K., and Gray, S.T., 2009, Ecology and the ratchet of events—Climate variability, niche dimensions, and species distributions: Proceedings of the National Academy of Sciences, v. 106, Supplement 2, p. 19,685–19,692, doi:10.1073/pnas.0901644106.

Jacob, T., Wahr, J., Pfeffer, W.T., and Swenson, S., 2012, Recent contributions of glaciers and ice caps to sea level rise: Nature, v. 482, no. 7386, p. 514-518, doi:10.1038/nature10847.

Janout, M.A., Weingartner, T.J., Royer, T., and Danielson, S., 2010, On the nature of winter cooling and the recent temperature shift on the northern Gulf of Alaska shelf: Journal of Geophysical Research, v. 115, C05023, 14 p., doi:10.1029/2009JC005774.

Janowicz, R., 2009, Recent hydrologic response trends observed in the Yukon Territory: IP3 Annual Science Meeting and Workshop, Lake Louise, Alberta, Canada, October 14–17, 2009.

Jay, C.V., Marcot, B.G., and Douglas, D.C., 2011, Projected status of the Pacific walrus (Odobenus rosmarus divergens) in the twenty-first century: Polar Biology, v. 34, p. 1,065–1,084.

Jenkerson, C.B., Maiersperger, T., and Schmidt, G., 2010, eMODIS–A user-friendly data source: U.S. Geological Survey Open-File Report 2010-1055, 10 p. (Also available at http://pubs.usgs.gov/of/2010/1055/.)

Jevrejeva, S., Moore, J.C., and Grinsted, A., 2010, How will sea level respond to changes in natural and anthropogenic forcings by 2100?: Geophysical Research Letters, v. 37, L07703, 5 p., doi:10.1029/2010GL042947.

Jia, G.J., Epstein, H.E., and Walker, D.A., 2003, Greening of arctic Alaska, 1981–2001: Geophysical Research Letters, v. 30, no. 20, p. 2,067, doi:10.1029/2003GL018268.

Johnstone, J., and Chapin F.S., III., 2006, Effects of soil burn severity on post–fire tree recruitment in boreal forest: Ecosystems, v. 9, no. 1, p. 14–31.

Johnstone, J.F., Chapin, F.S., III, Hollingsworth, T. N., Mack, M.C., Romanovsky, V., and Turetsky, M., 2010a, Fire, climate change, and forest resilience in Interior Alaska: Canadian Journal of Forest Research, v. 40, no. 7, p. 1,302–1,312.

Johnstone, J.F., Hollingsworth, T.N., Chapin, F.S., and Mack, M.C., 2010b, Changes in fire regime break the legacy lock on successional trajectories in Alaskan boreal forest: Global Change Biology, v. 16, no. 4, p. 1,281–1,295.

Johnstone, J.F., and Kasischke, E.S., 2005, Stand-level effects of soil burn severity on post-fire regeneration in a recently-burned black spruce forest: Canadian Journal of Forest Research, v. 35, no. 9, p. 2,151–2,163.

Johnstone, J.F., Rupp, T.S., Olson, M., and Verbyla, D., 2011, Modeling impacts of fire severity on successional trajectories and future behavior in Alaskan boreal forests: Landscape Ecology, v. 26, no. 4, p. 487–500.

Joint-Federal State Land Use Planning Commission, 1973, Major ecosystems of Alaska map: U.S. Geological Survey, scale 1:2,500,000.

Joly, K., Chapin, F.S., III, and Klein, D.R., 2010, Winter habitat selection by caribou in relation to lichen abundance, wildfires, grazing and landscape characteristics in northwest Alaska: Ecoscience, v. 17, no. 3, p. 321–333.

Jones, B.M., Arp, C.D., Beck, R.A., Grosse, G., Webster, J., and Urban, F.E., 2009c, Erosional history of Cape Halkett and contemporary monitoring of bluff retreat, Beaufort Sea coast, Alaska: Polar Geography, v. 32, p. 129–142.

Jones, B., Arp, C., Hinkel, K., Beck, R., Schmutz, J., Winston, B., 2009a, Arctic lake physical processes and regimes with implications for winter water availability and management in the National Petroleum Reserve Alaska: Environmental Management, v. 43, no. 6, p. 1,071–1,084.

Jones, B.M., Arp, C.D, Jorgenson, M.T., Hinkel, K.M., Schmutz, J.A., and Flint, P.L., 2009b, Increase in the rate and uniformity of coastline erosion in Arctic Alaska: Geophysical Research Letters, v. 36, L03502, 5 p., doi:10.1029/2008GL036205.

Jones, B.M., Hinkel, K.M., Arp, C.D., and Eisner, W.R., 2008, Modern erosion rates and loss of coastal features and sites, Beaufort Sea coastline, Alaska: Arctic, v. 61, no. 4, p. 361–372.

Jones, B.M., Grosse, G., Arp, C.D., Jones, M.C., Walter Anthony, K.M., and Romanovsky, V.E., 2011, Modern thermokarst lake dynamics in the continuous permafrost zone, northern Seward Peninsula, Alaska: Journal of Geophysical Research, v. 116, G00M03, 13 p., doi:10.1029/2011JG001666.

Jorgenson, M.T., Pullman, E.R., and Shur, Y.L., 2003, Geomorphology of the Northeast Planning Area of the National Petroleum Reserve-Alaska, 2002, second annual report: Anchorage, Alaska, Phillips Alaska, Inc., 67 p.

Jorgenson, M.T., Racine, C.H., Walters, J.C., and Osterkamp, T.E., 2001, Permafrost degradation and ecological changes associated with a warming climate in central Alaska: Climate Change, v. 48, p. 551–579.

Jorgenson, M.T., Romanovsky, V., Harden, J., Shur, Y., O'Donnell, J., Schuur, E.A.G., Kanevskiy, M., and Marchenko, S., 2010, Resilience and vulnerability of permafrost to climate change: Canadian Journal of Forest Research, v. 40, no. 7, p. 1,219–1,236.

Jorgenson, M.T., and Shur, Y., 2007, Evolution of lakes and basins in northern Alaska and discussion of the thaw lake cycle: Journal of Geophysical Research, v. 112, F02S17, 12 p., doi:10.1029/2006JF000531.

Jorgenson, M.T., Shur, Y.L., and Pullman, E.R., 2006, Abrupt increase in permafrost degradation in Arctic Alaska: Geophysical Research Letters, v. 33, L02503, 4 p., doi:10.1029/2005GL024960.

Jorgenson, T., Yoshikawa, K., Kanevskiy, M., Shur, Y., Romanovsky, V., Marchenko, S., Grosse, G., Brown, J., and Jones, B., 2008, Permafrost characteristics of Alaska: Ninth International Conference on Permafrost, University of Alaska Fairbanks–Fairbanks, p. 183–184.

Juday, G.P., Barber, V., Duffy, P., Linderholm, H., Rupp, T.S., Sparrow, S., Vaganov, E., and Yarie, J., 2005, Forests, land management, and agriculture, in Arctic Climate Impact Assessment: Cambridge, United Kingdom, and New York, Cambridge University Press, p.781–862, accessed June 8, 2012, at http://www.acia.uaf.edu/pages/scientific.html.

Jungclaus, J.H., Haak, H., Latif, M., and Mikolajewicz, U., 2005, Arctic-North Atlantic interactions and multidecadal variability of the meridional overturning circulation: Journal of Climate, v. 18, no. 19, p. 4,013–4,031.

Jutterström, S., and Anderson, L.G., 2005, The saturation of calcite and aragonite in the Arctic Ocean: Marine Chemistry, v. 94, p. 101–110.

Kane, D.L., Hinzman, L.D., McNamara, J.P., Zhang, Z., and Benson, C.S., 2000, An overview of a nested watershed study in Arctic Alaska: Nordic Hydrology, v. 31, no. 3–4, p. 245–266.

Kanevskiy, M., Shur, Y., Fortier, D., Jorgenson, M.T., and Stephani, E., 2011, Cryostratigraphy of late Pleistocene syngenetic permafrost (yedoma) in northern Alaska, Itkillik River exposure: Quaternary Research, v. 75, no. 3, p. 584–596.

Karl, T.R., Melillo, J.M., and Peterson, T.C., eds., 2009a, Global climate change impacts in the United States: New York, Cambridge University Press, 196 p.

Karl, T.R., Melillo, J.M., and Petersen, T.C., eds., 2009b, Regional climate impacts—Alaska, *in* Global climate change impacts in the United States: Cambridge, United Kingdom, Cambridge University Press, accessed June 8, 2011, at http://www.globalchange.gov/publications/reports/scientific-assessments/us-impacts.

Kaser, G., Cogley, J.G., Dyurgerov, M.B., Meier, M.F., and Ohmura, A., 2006, Mass balance of glaciers and ice caps—Consensus estimates for 1961–2004: Geophysical Research Letters, v. 33, L19501, 5 p., doi:10.1029/2006GL027511.

Kasischke, E.S., and Johnstone, J.F., 2005, Variation in ground-layer surface fuel consumption and its effects on site characteristics in a *Picea mariana* forest complex in Interior Alaska: Canadian Journal of Forest Research, v. 35, no. 9, p. 2,164–2,177.

Kasischke, E.S., and Turetsky, M.R., 2006, Recent changes in the fire regime across the North American boreal region—Spatial and temporal patterns of burning across Canada and Alaska: Geophysical Research Letters, v. 33, L09703, 5 p., doi:10.1029/2006GL025677.

Kasischke, E.S., Verbyla, D.L., Rupp, T.S., McGuire, A.D., Murphy, K.A., Jandt, R., Barnes, J.L., Hoy, E.E., Duffy, P.A., Calef, M., and Turetsky, M.R., 2010, Alaska's changing fire regime—implications for the vulnerability of its boreal forests: Canadian Journal of Forest Research, v. 40, no. 7, p. 1,313–1,324.

Kasper, J., and Weingartner, T.J., 2012, Modeling winter circulation under landfast ice—The interaction of winds with landfast ice: Journal of Geophysical Research, doi:10.1029/2011JC007649, in press.

Kattsov, V.M., Ryabinin, V.E., Overland, J.E., Serreze, M.C., Visbeck, M., Walsh, J.E., Meier, W., and Zhang, X., 2010, Arctic sea ice change–a grand challenge of climate science: Journal of Glaciology, v. 56, no. 200, p. 1,115–1,121, doi:10.3189/002214311796406176.

Kelly, B.P., Ainsworth, T.A., Boyce Jr., D.A., Hood, E., Murphy, P., and Powell, J., 2007, Climate change—Predicted impacts on Juneau: Report to Mayor Bruce Botelho and the City and Borough of Juneau Assembly, 86 p., accessed September 12, 2012, at http://www.juneau.org/clerk/boards/CBJ_Board_Archive_Page.php/.

Kenai Peninsula Borough, 2012, Spruce bark beetle mitigation program: website, accessed August 10, 2012, at http://www2.borough.kenai.ak.us/sbb/.

Khrustalev, L., 2001, Problems of permafrost engineering as related to global climate warming, *in* Paepe, R., and Melnikov, V.P., eds., Permafrost response on economic development, environmental security and natural resources: NATO Advanced workshop, Novosibirsk, Russia, November 12-16, 1998, Kluwer Academic Publishers, p. 407–423.

Kim, S.-J., Flato, G.M., and Boer, G.J., 2003, A coupled climate model simulation of the last glacial maximum, part 2—Approach to equilibrium: Climate Dynamics, v. 20, p. 635–661.

Kim, S.-J., Flato, G.M., Boer, G.J., and McFarlane, N.A., 2002, A coupled climate model simulation of the last glacial maximum, part 1—Transient multi–decadal response: Climate Dynamics, v. 19, p. 515–537.

Kinner, N.E., Lippmann, T., Ravens, T.M., and Zufelt, J.E., 2009, Implications of climate change and research needs for coastal processes in cold regions: Presented at the Climate Change Impacts on Defense Assets in Alaska Workshop, July 7-9, 2009, 24 p., accessed June 11, 2012, at http://defenseassets09.uaa.alaska.edu/CCICoastalProc070709.pdf.

Klein, E., Berg, E.E., and Dial, R., 2005, Wetland drying and succession across the Kenai Peninsula lowlands, south-central Alaska: Canadian Journal of Forest Research, v. 35, no. 8, p. 1,931–1,941.

Kofinas, G.P., Chapin, F.S., III, BurnSilver, S., Schmidt, J.I., Fresco, N.L, Kielland, K., Martin, S., Springsteen, A., and Rupp, T.S., 2010, Resilience of Athabascan subsistence systems to interior Alaska's changing climate: Canadian Journal of Forest Research, v. 40, no. 7, p. 1,347–1,359, doi:10.1139/X10-108.

Kokelj, S., Jenkins, R., and Milburn, D., 2005, The influence of thermokarst disturbance on the water quality of small upland lakes, Mackenzie Delta region, Northwest Territories: Canada: Permafrost and Periglacial Processes, v. 16, no. 4, p. 343–353.

Kunkel, K.E., Pielke, R.A, Jr., and Changnon, S.A., 1999, Temporal fluctuations in weather and climate extremes that cause economic and human health impacts—A review: Bulletin of American Meteorological Society, v. 80, p. 1,077–1,098.

Kurz, W.A., and Apps, M.J., 1999, A 70-year retrospective analysis of carbon fluxes in the Canadian forest sector: Ecological Applications, v. 9, p. 526–547.

Kwok, R., and Untersteiner, N., 2011, The thinning of Arctic sea ice: Physics Today, v. 64, no. 4, p. 36-41.

Laidre, K.L., Stirling, I., Lowry, L.F., Wiig, Ø., Heide-Jørgensen, M.P., and Ferguson, S.H., 2008, Quantifying the sensitivity of Arctic marine mammals to climate–induced habitat change: Ecological Applications, v. 18, no. S2, p. S97–S125.

Lapina, I., and Carlson, M.L., 2004, Non-native plant species of Susitna, Matanuska, and Copper River basins—Summary of survey findings and recommendations for control actions: Final Report for U.S. Department of Agriculture, U.S. Forest Service, State and Private Forestry, Anchorage, Alaska, accessed June 11, 2012, at http://www.uaa.alaska.edu/enri/publications/upload/Non-native_plants_final-report.pdf.

Larsen, C.F., Motyka, R.J., Arendt, A., Echelmeyer, K.A., and Geissler, P.E., 2007b, Glacier changes in southeast Alaska and northwest British Columbia and contribution to sea level rise: Journal of Geophysical Research, v. 112, F01007, 11 p., doi:10.1029/2006JF000586.

Larsen, C.F., Motyka, R.J., Freymueller, J.T., Echelmeyer, K.A., and. Ivins, E.R., 2005, Rapid viscoelastic uplift in southeast Alaska caused by post–Little Ice Age glacial retreat: Earth and Planetary Science Letters, v. 237, p. 548–560.

Larsen, P., and Goldsmith, S., 2007, How much might climate change add to future costs for public infrastructure in Alaska?: Institute of Social and Economic Research at the University of Alaska–Anchorage, Research Summary No. 8, June 2007, accessed June 11, 2012, at http://www.iser.uaa.alaska.edu/Publications/Juneclimatefinal.pdf.

Larsen, P.H., Goldsmith, S., Smith, O., Wilson, M.L., Strzepek, K., Chinowsky, P., and Saylor, B., 2007a, Estimating future costs for Alaska public infrastructure at risk from climate change: Institute of Social and Economic Research at the University of Alaska–Anchorage, project report, 108 p., accessed June 11, 2012, at http://www.iser.uaa.alaska.edu/Publications/JuneICICLE.pdf.

Larsen, P.H., Goldsmith, S., Smith, O., Wilson, M.L., Strzepek, K., Chinowsky, P., and Saylor, B., 2008, Estimating future costs for Alaska public infrastructure at risk from climate change: Global Environmental Change, v. 18, no. 3, p. 442–457.

Lawler, J.J., Tear, T.H., Pyke, C., Shaw. M.R., Gonzalez, P., Kareiva, P., Hansen, L., Hannah, L., Klausmeyer, K., Aldous, A., Cienz, C., and Pearsall, S., 2010, Resource management in a changing and uncertain climate: Frontiers in Ecology and the Environment, v. 8, no. 1, p. 35–43, doi:10.1890/070146.

Leask, L., Killorin, M., and Martin, S., 2001, Trends in Alaska's people and economy: Institute of Social and Economic Research at the University of Alaska–Anchorage, 16 p., accessed June 11, 2012, at http://www.iser.uaa.alaska.edu/Publications/Alaska2020.pdf.

Lemelin, H., Dawson, J., Stewart, E J., Maher, P. and Lueck, M., 2010, Last-chance tourism—The boom, doom, and gloom of visiting vanishing destinations: Current Issues in Tourism, v. 13, no. 5, p. 477–493.

Lemke, P., Ren, J., Alley, R.B., Allison, I., Carrasco, J., Flato, G., Fujii, Y., Kaser, G., Mote, P., Thomas, R.H., and Zhang, T., 2007, Observations—Changes in snow, ice, and frozen ground, in Solomon, S., Qin, D., Manning, M., Chen, Z., Marquis, M., Averyt, K.B., Tignor, M., and Miller, H.L., eds., Climate change 2007—The physical science basis—Contribution of Working Group I to the Fourth Assessment Report of the Intergovernmental Panel on Climate Change: Cambridge, United Kingdom, and New York, Cambridge University Press, accessed July 20, 2012, at http://www.ipcc.ch/publications_and_data/ar4/wg1/en/ch4.html.

Lewis, P.N., Hewitt, M.J., Riddle, C.L., and McMinn, A., 2003, Marine introductions in the Southern Ocean—An unrecognized hazard to biodiversity: Marine Pollution Bulletin, v. 46, no. 2, p. 213–223.

Lewis, P.N., Riddle, M.J., and Hewitt, C.L., 2004, Management of exogenous threats to Antarctica and the sub-Antarctic Islands—Balancing risks from TBT and non-indigenous marine organisms: Marine Pollution Bulletin, v. 49, no. 11–12, p. 999–1,005.

L'Heureux, M., Kumar, A., Bell, G., Halpert, M., and Higgins, R., 2008, Role of the Pacific-North American (PNA) pattern in the 2007 Arctic sea ice decline: Geophysical Research Letters, v. 35, L20701, 5 p., doi:10.1029/2008GL035205.

Li, W.K.W., McLaughlin, F.A., Lovejoy, C., and Carmack, E.C., 2009, Smallest algae thrives as the Arctic Ocean freshens: Science, v. 326, no. 5952, p. 539, doi:10.1126/science.1179798.

Liston, G.E., and Hiemstra, C.A., 2011, The changing cryosphere—Pan-Arctic snow trends (1979–2009): Journal of Climate, v. 24, no. 21, p. 5,691–5,712, doi:10.1175/JCLI-D-11-00081.1.

Loring, P.A., and Gerlach, S.C., 2009, Food, culture, and human health in Alaska—An integrative health approach to food security: Environmental Science and Policy, v. 12, no. 4, p. 466–478.

Loring, P.A., and Gerlach, S.C., 2010, Food security and conservation of Yukon River salmon—Are we asking too much of the Yukon River?: Sustainability, v. 2, no. 9, p. 2,965–2,987, doi:10.3390/su2092965.

Loring, P.A., Duffy, L.K., and Murray, M.S., 2010, A risk-benefit analysis of wild fish consumption for various species in Alaska reveals shortcomings in data and monitoring needs: Science of The Total Environment, v. 408, no. 20, p. 4,532–4,541.

Loring, P.A., Gerlach, S.C., Atkinson, D.E., and Murray, M.S., 2011, Ways to help and ways to hinder—Governance for successful livelihoods in a changing climate: Arctic, v. 64, no. 1, p. 73–88.

Lovvorn, J.R., Grebmeier, J.M., Cooper, L.W., Bump, J.K., and Richman, J.G., 2009, Modeling marine protected areas for threatened eiders in a climatically shifting Bering Sea: Ecological Applications, v. 19, no. 6, p. 1,596–1,613.

Lynch, J.A., Clark, J.S., Bigelow, N.H., Edwards, M.E., and Finney, B.P., 2002, Geographic and temporal variations in fire history in boreal ecosystems of Alaska: Journal of Geophysical Research, v. 108, 8152, 17 p., doi:10.1029/2001JD000332.

Macdonald, R., Harner, T., and Fyfe, J., 2005, Recent climate change in the Arctic and its impact on contaminant pathways and interpretation of temporal trend data: Science of the Total Environment, v. 342, no. 1, p. 5–86.

Mack, M.C., Bret-Harte, M.S., Hollingsworth, T.N., Jandt, R.R., Shuur, E.A.G, Shaver, G.R., and Verbyla, D.L., 2011, Carbon loss from an unprecedented Arctic tundra wildfire: Nature, v. 475, no. 7357, p. 489–492.

Maier, J.A.K., Ver Hoef, J., McGuire, A.D., Bowyer, R.T., Saperstein, L., and Maier, H.A., 2005, Distribution and density of moose in relation to landscape characteristics— Effects of scale: Canadian Journal of Forest Research, v. 35, no. 9, p. 2,233–2,243.

Mantua, N.J., Hare, S.R., Zhang, Y., Wallace, J.M., and Francis, R.C., 1997, A Pacific decadal climate oscillation with impacts on salmon: Bulletin of the American Meteorological Society, v. 78, p. 1,069–1,079.

Marasco, R.J., Goodman, D., Grimes, C.B., Lawson, P.W., Punt, A.E., and Quinn, T.J., II, 2007, Ecosystem-based fisheries management—Some practical suggestions: Canadian Journal of Fisheries and Aquatic Sciences, v. 64, no. 6, p. 928–939.

Marchenko, S., Romanovsky, V., and Tipenko, G., 2008, Numerical modeling of spatial permafrost dynamics in Alaska, in Kane, D.L., and Hinkel, K.M., eds., Proceedings of the Ninth International Conference on Permafrost, June 29–July 3, 2008, Fairbanks, Alaska: Institute of Northern Engineering at the University of Alaska– Fairbanks, v. 2, p. 1,125–1,130.

Markon, C.J., 1992, Land cover mapping of the Upper Kuskokwim Resource Management Area, using Landsat and digital data base approach: Canadian Journal of Remote Sensing, v. 18, no. 2, p. 62–71.

Markon, C.J., 1995, The history and use of remote sensing for conservation and management of federal lands in Alaska: Natural Areas Journal, v. 15, p. 329–338.

Markon, C.J., and Wesser, S., 1996, Land cover mapping of the National Park Service Northwest Alaska Management Area using Landsat multispectral and thematic mapper satellite data: National Resource Technical Report NPS/ARCN/ NRTR-1996-001, National Park Service, Fort Collins, Colo.

Mars, J.C., and Houseknecht, D.W., 2007, Quantitative remote sensing study indicates doubling of coastal erosion rate in past 50 years along a segment of the Arctic coast of Alaska: Geology, v. 35, no. 7, p. 583–586.

Martin, P.D., Jenkins, J.L., Adams, F.J., Jorgenson, M.T., Matz, A.C., Payer, D.C., Reynolds, P.E., Tidwell, A.C., and Zelenak, J.R., 2009, Wildlife response to environmental Arctic change—Predicting future habitats of Arctic Alaska: U.S. Fish and Wildlife Service, Report from the Wildlife Response to Environmental Arctic Change (WildREACH)— Predicting Future Habitats of Arctic Alaska Workshop, November 17–18, 2008, Fairbanks, Alaska, 138 p., accessed June 11, 2012, at http://www.arcus.org/alaskafws/ downloads/pdf/WildREACH_Workshop_Report_Final.pdf.

Maslanik, J., Fowler, C., Stroeve, J., Drobot, S., Zwally, J., Yi, D. and Emery, W., 2007, A younger, thinner Arctic ice cover—Increased potential for rapid, extensive sea-ice loss: Geophysical Research Letters, v. 34, L24501, 5 p., doi:10.1029/2007GL032043.

Maslanik, J., Stroeve, J., Fowler, C., and Emery, W., 2011, Distribution and trends in Arctic sea ice age through spring 2011: Geophysical Research Letters, v. 38, L13502, 6 p., doi:10.1029/2011GL047735.

Maslowski, W, Kinney, J., Jakacki, J., and Zwally, J., 2008, State of the Arctic sea ice: Geophysical Research Abstracts, v. 10, EGU2008-A-06425.

Mathis, J.T., Cross, J.N., and Bates, N.R., 2011, Coupling primary production and terrestrial runoff to ocean acidification and carbonate mineral suppression in the eastern Bering Sea: Journal of Geophysical Research, v. 116, C02030, 24 p., doi:10.1029/2010JC006453.

Maunder, M.N., and Watters, G.M., 2003, A-SCALA—an age structured statistical catch-at-length analysis for assessing tuna stocks in the eastern Pacific Ocean: Inter-American Tropical Tuna Commission Bulletin, v. 22, no. 5, p. 433–582.

McDowell Group, 2010, Economic impact of Alaska's visitor industry—March 2010: Prepared for the State of Alaska Department of Commerce, Community, and Economic Development, Division of Economic Development, accessed June 11, 2012, at http://www.commerce.state.ak.us/ded/dev/ pub/Visitor_Industry_Impacts_3_30.pdf.

McDowell Group, 2011a, The role of the oil and gas industry in Alaska's economy: Prepared for Alaska Oil and Gas Association, 55 p.

McDowell Group, 2011b, Alaska visitor statistics program VI interim visitor volume report—Fall/Winter 2010–2011: Prepared for the State of Alaska Department of Commerce, Community, and Economic Development, Division of Economic Development, accessed June 11, 2012, at http://www.dced.state.ak.us/ded/dev/toubus/pub/AVSP_VI_Fall_Winter_2010_2011_REVISED.pdf.

McDowell Group, 2011c, Alaska visitor statistics program VI interim visitor volume report—Summer 2010: Prepared for the State of Alaska Department of Commerce, Community, and Economic Development, Division of Economic Development, accessed June 11, 2012, at http://www.dced.state.ak.us/ded/dev/toubus/pub/AVSP_VI_Summer_2010_REVISED.pdf.

McGuire, A.D., Ruess, R., Lloyd, A., Yarie, J., Clein, J., and Juday, G., 2010, Vulnerability of white spruce tree growth in interior Alaska in response to climate variability—Dendrochrological demographic, and experimental perspectives: Canadian Journal of Forest Research, v. 40, p. 1,197–1,209, doi:10.1139/X09-206.

McLaughlin, J., DePaola, A., Bopp, C., Martinek, K., Napolilli, N., Allison, C., Murray, S., Thompson, E., Bird, M., and Middaugh, J., 2005, Outbreak of *Vibrio parahaemolyticus* gastroenteritis associated with Alaskan oysters: New England Journal of Medicine, v. 353, no. 14, p. 1,463–70.

McNeeley, S., 2009, Seasons out of balance—Vulnerability and sustainable adaptation in Alaska: University of Alaska–Fairbanks, Ph.D. dissertation, accessed July 23, 2012, at http://www.uaf.edu/files/rap/McNeeley_Dissertation_2009.pdf.

McNeeley, S., 2012, Alaska—Examining barriers and opportunities for sustainable adaptation to climate change in interior Alaska: Climate Change, v. 111, no. 3, p. 835–857.

McNeeley, S., and Huntington, O., 2007, Postcards from the (not so) frozen north—Talking about climate change in Alaska, *in* Creating a climate for change: Cambridge, United Kingdom, Cambridge University Press, p. 139–152.

McNeeley, M., and Shulski, M., 2011, Anatomy of a closing window—Vulnerability to changing seasonality in interior Alaska: Global Environmental Change, v. 21, no. 2, p. 464–473, doi:10.1016/j.gloenvcha.2011.02.003.

Meier, W., Stroeve, J., and Fetterer, F., 2007, Whither Arctic sea ice?—A clear signal of decline regionally, seasonally, and extending beyond the satellite record: Annals of Glaciology, v. 46, no. 1, p. 428–434.

Melsom, A., Metzger, E.J., and Hurlburt, H.E., 2003, Impact of remote oceanic forcing on Gulf of Alaska sea levels and mesoscale circulation: Journal of Geophysical Research, v. 108, no. 3346, 16 p., doi:10.1029/2002JC001742.

Mendez, J., Hinzman, L.D., and Kane, D.L., 1998, Evapotranspiration from a wetland complex on the arctic coastal plain of Alaska: Nordic Hydrology, v. 29, no. 4–5, p. 303–330.

Mills, J., Hunt, J., Rosenberger, A., Daum, D., and Solie, D., 2008, EPSCoR climate change research—Examination of long-term ice break-up and freeze–up trends in the Yukon River based on historical data sources collected for the Yukon River temperature archive project, *in* Growing sustainability science in the north—Science, policy, education, legacy in the international polar year: Fairbanks, Alaska, American Association for the Advancement of Science, Arctic Division, p. 41, accessed June 11, 2012, at http://arcticaaas.org/meetings/2008/2008_aaas_abstracts_v1.4.pdf.

Mills, P., 1994, The agriculture potential of northwestern Canada and Alaska and the impact of climate change: Arctic, v. 47, no. 2, p. 115–123.

Moerlein, K.J., and Carothers, C., 2012, Total environment of change—Impacts of climate change and social transitions on subsistence fisheries in northwest Alaska: Ecology and Society, v. 17, no. 1, 10 p.

Moore, S.E., and Huntington, H.P., 2008, Arctic marine mammals and climate change—Impacts and resilience: Ecological Applications, v. 18, no. sp2, p. 157–165.

Morita, K., and Fukawaka, M., 2007, Why age and size at maturity have changed in Pacific salmon: Marine Ecology Progress Series, v. 335, p. 289–294.

Morrison, J., Foreman, M.G.G., and Masson, D., 2011, A method for estimating monthly freshwater discharge affecting British Columbia coastal waters: Atmosphere–Ocean, v. 50, no. 1, p. 1–8, doi:10.1080/07055900.2011.637667.

Motyka, R.J., Fahnestock, M., and Truffer, M., 2010, Volume change of Jakobshavn Isbræ, West Greenland—1985–1997–2007: Journal of Glaciology, v. 56, n. 198, p. 635–646, doi:10.3189/002214310793146304.

Motyka, R.J., and Truffer, M., 2007, Hubbard Glacier, Alaska—2002 closure and outburst of Russell Fjord and postflood conditions at Gilbert Point: Journal of Geophysical Research, v. 112, F02004, 15 p., doi:10.1029/2006JF000475.

Mueter, F.J., Boldt, J.L., Megrey, B.A., and Peterman, R.M., 2007, Recruitment and survival of northeast Pacific Ocean fish stocks—Temporal trends, covariation, and regime shifts: Canadian Journal of Fisheries and Aquatic Sciences, v. 64, no. 6, p. 911–992.

Mueter, F.J., and Litzow, L.A., 2008, Sea ice retreat alters the biogeography of the Bering Sea continental shelf: Ecological Applications, v. 18, p. 309–320.

Mueter, F.J., Peterman, R.M., and Pyper, B.J., 2002, Opposite effects of ocean temperature on survival rates of 120 stocks of Pacific salmon (*Onchorynchus* spp.) in northern and southern areas: Canadian Journal of Fisheries and Aquatic Science, v. 59, p. 456–463.

Mundy, P.R., 2005, The Gulf of Alaska—Biology and oceanography: Alaska Sea Grant College Program, University of Alaska–Fairbanks, 214 p., doi:10.4027/gabo.2005.

Mundy, P.R., and Evenson, D.F., 2011, Environmental controls of phenology of high-latitude Chinook salmon populations of the Yukon River, North America, with application to fishery management: Journal of Marine Science, v. 68, no. 6, p. 1,155–1,164, doi:10.1093/icesjms/fsr080.

Murphy, K., Huettmann, F., Fresco, N., and Morton, J., 2010, Connecting Alaska landscape into the future: Final report to the U.S. Fish and Wildlife Service, August 2010, 98 p.

Myers, K., Walker, R., Davis, N., Armstrong, J., Fourrier, W., Mantua, N., and Raymond-Yakoubian, J., 2010, 2010 Arctic Yukon Kuskokwim (AYK) sustainable salmon initiative project final product—Climate–ocean effects on YAK Chinook salmon: Arctic-Yukon-Kuskokwim Sustainable Salmon Initiative, 267 p.

Nakicenovic, G., Alcamo, J., Davis, G., deVries, B., Fenhann, J., Gaffin, S., Gregory, K., Grubler, A., Jung, T.Y., Kram, T., La Rovere, E., Michaelis, L., Mori, S., Morita, T., Pepper, W., Pitcher, H., Price, L., Riahi, K., Roehrl, A., Rogner, H., Sankovski, A., Schlesinger, A., Shukla, P., Smith, S., Swart, R., Van Rooijen, S., Victor, N., and Zhou, D., 2000, Special report on emissions scenarios: Cambridge, United Kingdom, and New York, Cambridge University Press, 570 p., accessed July 24, 2012, at http://www.ipcc.ch/ipccreports/sres/emission/index.php?idp=0

National Assessment Synthesis Team, 2000, Climate change impacts on the United States—The potential consequences of climate variability and change: U.S. Global Change Research Program, United Kingdom, Cambridge University Press, accessed June 11, 2012, at http://www.gcrio.org/NationalAssessment/index.htm.

National Assessment Synthesis Team, 2003, United States national assessment of the potential consequences of climate variability and change educational resources regional paper—Alaska: National Assessment Synthesis Team, accessed June 11, 2012, at http://www.usgcrp.gov/usgcrp/nacc/education/alaska/default.htm.

National Marine Fisheries Service, 1999, Ecosystem-based fishery management—A report to Congress by the Ecosystem Principles Advisory Panel: National Marine Fisheries Service, 54 p., accessed June 11, 2012, at http://www.nmfs.noaa.gov/sfa/EPAPrpt.pdf.

National Marine Fisheries Service, 2010, Fisheries of the United States—2009: National Marine Fisheries Service web page, last modified September 3, 2010, accessed June 11, 2012, at http://www.st.nmfs.noaa.gov/st1/fus/fus09/index.html.

National Oceanic and Atmospheric Administration, 2010, Bering climate and ecosystem—Bering Sea status and overview: National Oceanic and Atmospheric Administration website, accessed June 11, 2012, at http://www.beringclimate.noaa.gov/bering_status_overview.html.

National Oceanic and Atmospheric Administration, 2011, Arctic Report Card—Update for 2011, website, accessed August 10, 2012, at http://www.arctic.noaa.gov/reportcard/index.html.

National Oceanic and Atmospheric Administration, 2012a, Distributed Biological Observatory, website, accessed August 14, 2012 at http://www.arctic.noaa.gov/dbo/.

National Oceanic and Atmospheric Administration, 2012b, Climate Variability and Marine Fisheries: PFEL Climate and Marine Fisheries, website, accessed August 10, 2012, at http://www.pfeg.noaa.gov/research/climatemarine/cmffish/cmffishery6.html.

National Oceanic and Atmospheric Administration, 2012c, National Weather Service - Climate Prediction Center, website, accessed August 10, 2012, at http://www.cpc.ncep.noaa.gov/.

National Oceanic and Atmospheric Administration, 2012d, North Atlantic Oscillation: National Weather Service – Climate Prediction Center, website, accessed September 12, 2012, at http://www.cpc.ncep.noaa.gov/data/teledoc/nao.shtml.

National Oceanic and Atmospheric Administration, 2012e, Pacific Decadal Oscillation: National Climatic Data Center, website, accessed September 12, 2012, at http://www.ncdc.noaa.gov/teleconnections/pdo/.

National Park Service, 2012a, Inventory and Monitoring Program – Network websites, website, accessed August 10, 2012, at http://science.nature.nps.gov/im/networks.cfm.

National Park Service, 2012b, "Rehearsing the Future," Scenario Planning in Alaska, website, accessed August 10, 2012, at http://www.nps.gov/akso/nature/climate/scenario.cfm.

National Phenology Network, 2012, website, accessed August 10, 2012, at http://www.usanpn.org/.

National Research Council, 2003, Cumulative environmental effects of oil and gas activities on Alaska's North Slope: Washington, D.C., National Academies Press, 304 p., accessed June 11, 2012, at http://www.nap.edu/catalog.php?record_id=10639.

National Research Council, 2010, Ocean Acidification—A national strategy to meet the challenges of a changing ocean: Washington, D.C., National Academies Press.

National Snow and Ice Data Center, 2012, website, accessed August 10, 2012, at http://nsidc.org/.

Neal, E.G., Hood, E., and Smikrud, K., 2010, Contribution of glacier runoff to freshwater discharge into the Gulf of Alaska, Geophysical Research Letters, v. 37, no. 6, p. 1–5, doi:10.1029/2010GL042385.

Nelson, F.E., Anisimov, O.A., and Shiklomanov, N.I., 2001, Subsidence risk from thawing permafrost—The threat to man-made structures across regions in the far north can be monitored, Nature, v. 410, no. 6831, p. 889–890.

Nelson, J.L., Zavaleta, E.S., and Chapin, F.S., III, 2008, Boreal effects on subsistence resources in Alaska and adjacent Canada: Ecosystems, v. 11, no. 1, p. 156–171.

Nelson, R.K., 1969, Hunters of the northern ice: Chicago, Ill., University of Chicago Press.

Nelson, R.K., 1986, Hunters of the northern forest—Designs for survival among the Alaska Kutchin: Chicago, Ill., University of Chicago Press.

Nenana Ice Classic, 2012: Nenana Ice Classic website, accessed June 11, 2012, at http://www.nenanaakiceclassic.com/Breakup%20Log.html.

Newsome, S.D., Etnier, M.A., Kurle, C.M., Waldbauer, J.R., Chamberlain, C.P., and Koch, P.L., 2007, Historic decline in primary productivity in western Gulf of Alaska and eastern Bering Sea—Isotopic analysis of northern fur seal teeth: Marine Ecology Progress Series, v. 332, p. 211–224, doi:10.3354/meps332211.

Nghiem, S., Rigor, I., Perovich, D., Clemente-Colon, P., Weatherly, J., and Neumann, G., 2007, Rapid reduction of Arctic perennial sea ice: Geophysical Research Letters, v. 34, L19504, 6 p., doi:10.1029/2007GL031138.

Niebauer, H.J., 1998, Variability in Bering Sea ice cover as affected by a regime shift in the North Pacific in the period 1947–1996: Journal of Geophysical Research, v. 103, no. C12, p. 717–737, doi:10.1029/98JC02499.

Norris, F., 2002, Alaska subsistence: National Park Service, Anchorage, Alaska, accessed November 15, 2011, at http://www.nps.gov/history/history/online_books/norris1/index.htm.

North Pacific Fishery Management Council, 2009, Fishery management plan for fishery resources of the Arctic Management Area: Anchorage, Alaska, North Pacific Fishery Management Council, 146 p., accessed June 11, 2012, at http://www.fakr.noaa.gov/npfmc/PDFdocuments/fmp/Arctic/ArcticFMP.pdf.

North Pacific Fishery Management Council, 2010, Overview of the Aleutian Islands fishery ecosystem plan: Anchorage, Alaska, North Pacific Fishery Management Council, 22 p., accessed June 11, 2012, at http://www.pcouncil.org/wp-content/uploads/D1a_ATT1_ALEUTIAN_EFMP_NOV2010BB.pdf.

North Pacific Fishery Management Council, 2011, Groundfish species profiles: North Pacific Fishery Management Council, Anchorage, Alaska, 57 p., accessed June 11, 2012, at http://www.fakr.noaa.gov/npfmc/PDFdocuments/resources/Species_Profiles2011.pdf.

North Pacific Marine Science Organization (PICES), 2012, website, accessed August 10, 2012, at http://www.pices.int/.

North Slope Science Initiative, 2012, website, accessed August 10, 2012, at http://www.northslope.org/.

Northern Economics, 2004, Alaska visitor arrivals—Fall-Winter 2003–2004: Prepared for the State of Alaska, Department of Commerce, Community and Economic Development, p. 31, accessed June 11, 2012, at http://www.commerce.state.ak.us/ded/dev/toubus/pub/Fall_Winter_2003_04_arrivals.pdf.

Northern Economics, 2009, Briefings—Economic impacts of Alaska's fishing industry: Anchorage, Alaska, Northern Economics, Inc., v. 11, no. 1, 2 p., accessed July 20, 2012, at http://www.northerneconomics.com/relevance/news/newspdfs/april2009.pdf.

Northern Economics, 2011, Seafood industry in Alaska's economy: Juneau, Alaska, Northern Economics, Inc., 6 p., accessed June 11, 2012, at http://www.marineconservationalliance.org/wp-content/uploads/2011/02/SIAE_Feb2011a.pdf.

Nowacki, G., and Brock, T., 1995, Ecoregions and subregions of Alaska: U.S. Department of Agriculture, U.S. Forest Service, EcoMap Version 2.0, scale 1:5,000,000.

Nowacki, G., Spencer, P., Fleming, M., Brock, T., and Jorgenson, T., 2001, Unified ecoregions of Alaska—2001: U.S. Geological Survey, Open-File Report 2002–297, scale 1:2,500,000. (Also available at http://pubs.er.usgs.gov/publication/ofr2002297.)

O'Donnell, J.A., Aiken, G.R., Kane, E.S., and Jones, J.B., 2010, Source water controls on the character and origin of dissolved organic matter in streams of the Yukon River basin, Alaska: Journal of Geophysical Research, v. 115, G03025, 12 p., doi:10.1029/2009JG001153.

Oechel, W.C., Vourlitis, G.L., Hastings, S.J., Zulueta, R.C., Hinzman, L.D., and Kane, D.L., 2000, Acclimation of ecosystem CO_2 exchange in the Alaskan Arctic in response to decadal climate warming: Nature, v. 406, p. 978–981, doi:10.1038/35023137.

Oerlemans, J., 2005, Extracting a climate signal from 169 glacier records: Science, v. 308, no. 5722, p. 675-677, doi:10.1126/science.1107046.

Ogi, M., Yamazaki, K., and Wallace, J., 2010, Influence of winter and summer surface wind anomalies on summer Arctic sea ice extent: Geophysical Research Letters, v. 37, L07701, 5 p., doi:10.1029/2009GL042356.

Omernik, J.M., 1987, Ecoregions of the conterminous United States: Annals of the Association of American Geographers, v. 77, no. 1, p. 118–125.

Omernik, J.M., 1995, Ecoregions: A framework for managing ecosystems: The George Wright Forum, v. 12, no. 1, p. 35–51.

Omernik, J.M., Hughes, R.M., Griffith, G.E., and Hellyer, G., 2011, Common geographic frameworks, in Landscape and predictive tools—A guide to spatial analysis for environmental assessment: U.S. Environmental Protection Agency, National Center for Environmental Assessment Report EPA/100/R-11/002, ch. 5, Risk Assessment Forum, Washington, D.C., accessed June 11, 2012, at ftp://ftp.epa.gov/wed/ecoregions/pubs/Omernik_et_al_CommonGeogFrameworks_FinalSept2011.pdf.

Oregon State University, 2012, Parameter-elevation Regressions on Independent Slopes Model Climate Group, website, accessed August 10, 2012, at http://www.prism.oregonstate.edu/.

Orr, J.C., Fabry, V.J., Aumont, O., Bopp, L., Doney, S.C., Feely, R.A., Gnanadesikan, A., Gruber, N., Ishida, A., Joos, F., Key, R.M., Lindsay, K., Maier-Reimer, E., Matear, R., Monfray, P., Mouchet, A., Najjar, R.G., Plattner, G.K., Rodgers, K.B., Sabine, C.L., Sarmiento, J.L., Schlitzer, R., Slater, R.D., Totterdell, I J., Weirig, M.F., Yamanaka, Y., and Yool, A., 2005, Anthropogenic ocean acidification over the twenty-first century and its impact on calcifying organisms: Nature, v. 437, no. 7059, p. 681–686.

Osterkamp, T.E., 2003, A thermal history of permafrost in Alaska, in Proceedings of the Eighth International Conference on Permafrost, July 21–25, 2003: Zurich, Switzerland, Balkema Publishers, p. 863–868.

Osterkamp, T.E., 2008, Thermal state of permafrost in Alaska during the fourth quarter of the twentieth century (plenary paper), in Proceedings of the Ninth International Conference on Permafrost, June 29–July 3, 2008: Fairbanks, Alaska: v. 2, p. 1,333–1,338.

Overland, J.E., 2009, The case for global warming in the Arctic, in Nihoul, J.C.J., and Kostianoy, A.G., eds., Influence of climate change on the changing Arctic and sub-Arctic conditions: Dordrecht, Netherlands, Springer, p. 13–23.

Overland, J., Bhatt, U., Key, J., Liu, Y., Walsh, J., and Wang, M., 2011, Temperature and clouds, in Arctic Report Card—Update for 2011: National Oceanic and Atmospheric Administration website accessed June 11, 2012, at http://www.arctic.noaa.gov/reportcard/temperature_clouds.html.

Overland, J., Turner, J., Francis, J., Gillet, N., Marshall, G., and Tjernström, M., 2008, The Arctic and Antarctic—two faces of climate change: EOS, v. 89, no. 19, p. 177.

Overland, J.E., and Wang, M., 2007, Future regional Arctic sea ice declines: Geophysical Research Letters, v. 34, L17705, 7 p., doi:10.1029/2007GL030808.

Overland, J., and Wang, M., 2010, Large-scale atmospheric circulation changes associated with the recent loss of Arctic sea ice: Tellus, v. 62A, no. 1, p. 1–9.

Pacific Arctic Group, 2012, website, accessed August 14, 2012, at http://pag.arcticportal.org/.

Parkinson A., 2010, Sustainable development, climate change and human health in the Arctic: International Journal of Circumpolar Health, v. 69, no. 1, p. 100.

Parmesan, C., and Yohe, G., 2003, A globally coherent fingerprint of climate change impacts across natural system: Nature, v. 421, p. 37–42, doi:10.1038/nature01286.

Parson, E.A., Carter, L., Anderson, P., Wang, B., and Weller, G., 2001, Potential consequences and change for Alaska, in Climate change impacts on the United States: Cambridge, United Kingdom, Cambridge University Press, foundation report, p. 283–312.

Patrick, W.S., Spencer, P., Link, J., Cope, J., Field, J., Kobayashi, D., Lawson, P.W., Gedamke, T., Cortes, E., Ormseth, O., Bigelow, K., and Overholtz, W., 2010, Using productivity and susceptibility indices to assess the vulnerability of United States fish stocks to overfishing: Fishery Bulletin, v. 108, no. 3, p. 305–322.

Pavelsky, T.M., and Smith, L.C., 2004, Spatial and temporal patterns in Arctic river ice breakup observed with MODIS and AVHRR time series: Remote Sensing Environment, v. 93, p. 328–338.

Peacock, S., 2012, Projected 21st century changes in temperature, precipitation, and snow cover over North America in CCSM4: Journal of Climate, v. 25, p. 4405-4429, DOI: 10.1175/JCLI-D-11-00214.1.

Perry, A.L., Low, P.L., Ellis, J.R., and Reynolds, J.D., 2005, Climate change and distribution shifts in marine fishes: Science, v. 308, no. 5730, p.1,912–1,915, doi:10.1126/science.1111322.

Petersen, M.R., and Douglas, D.C., 2004, Winter ecology of spectacled eiders—Environmental characteristics and population change: Condor, v. 106, p. 79–94.

Pettersen, T., 2011, Putin sees bright future for Arctic transport: Barents Observer, September 25, 2011, accessed June 11, 2012, at http://barentsobserver.com/en/arkhangelsk-obl/putin-sees-bright-future-arctic-transport.

Pikitch, E.K., Santora, C., Babcock, E.A., Bakun, A., Bonfil, R., Conover, D.O., Dayton, P., Doukakis, P., Fluharty, D., Heneman, B., Houde, E.D., Link, J., Livingston, P.A., Mangel, M., McAllister, M.K., Pope, J., and Sainsbury, K.J., 2004, Ecosystem-based fishery management: Science, v. 305, no. 5682, p. 346–347, doi:10.1126/science.1098222.

Ping, C.-L., Michaelson, G.J., Jorgenson, M.T., Kimble, J.M., Epstein, H., Romanovsky, V.E., and Walker, D.A., 2008, High stocks of soil organic carbon in the North American Arctic region: Nature Geoscience, v. 1, no. 9, p. 615–619.

Ping, C.-L., Michaelson, G.J., Guo, L., Jorgenson, M.T., Kanevskiy, M., Shur, Y., Dou, F., and Liang, J., 2011, Soil carbon and material fluxes across the eroding Alaska Beaufort Sea coastline: Journal of Geophysical Research, v. 116, G02004, 12 p., doi:10.1029/2010JG001588.

Pojar, J., and MacKinnon, A., 1994, Plants of the Pacific Northwest Coast - Washington, Oregon, British Columbia, and Alaska: Redmond, Washington, Lone Pine Publishing.

Pope, V., Gallani, M.L., Rowntree, P.R., and Stratton R.A., 2000, The impact of new physical parameterizations in the Hadley Centre climate model—HadAM3: Climate Dynamics, v. 16, p. 123–146.

Post, A., O'Neel, S., Motyka, R. J., and Streveler, G., 2011, A complex relationship between calving glaciers and climate: EOS, Transactions American Geophysical Union, v. 92, no. 37, p. 305doi:10.1029/2011EO370001.

Pritchard, H.D., Luthcke, S.B., and Fleming, A.H., 2011, Understanding ice-sheet mass balance—Progress in satellite altimetry and gravimetry: Journal of Glaciology, v. 56, no. 200, p. 1,151–1,161, doi:10.3189/002214311796406194.

Prowse, T.D., and Beltaos, S., 2002, Climatic control of river-ice hydrology—A review: Hydrological Processes, v. 16, p. 805–822.

Public Law 101-606: Global Change Research Act of 1990 (104 Stat. 3096-3104, Date: November 16, 1990): Text from United States Public Laws, accessed October 30, 2012, at http://www.gcrio.org/gcact1990.html.

Public Law 109-58: Energy Policy Act of 2005 (119 Stat. 594-1143, Date: August 8, 2005): Test from United States Public Laws, accessed July 30, 2012, at http://www.gpo.gov/fdsys/pkg/BILLS-109hr6enr/pdf/BILLS-109hr6enr.pdf

Quadrelli, R., and Wallace, J., 2004, A simplified linear framework for interpreting patterns of Northern Hemisphere wintertime climate variability: Journal of Climate, v. 17, p. 3,728–3,744.

Randall, D.A., Wood, R.A., Bony, S., Colman, R., Fichefet, T., Fyfe, J., Kattsov, A., Pitman, A., Shukla, J., Srinivasan, J., Stougger, J., Sumi, A., and Taylor, K.E., 2007, Climate models and their evaluation, in Solomon, S., Qin, D., Manning, M., Chen, Z., Marquis, M., Averyt, K.B., Tignor, M., and Miller, H.L., eds., Climate change 2007—The physical science basis, contribution of working group I to the Forth Assessment Report of the Intergovernmental Panel on Climate Change: Cambridge, United Kingdom, and New York, Cambridge University Press, p. 589–662.

Rattenbury, K., Kielland, K., Finstad, G., and Schneider, W., 2009, A reindeer herder's perspective on caribou, weather, and socio-economic change on the Seward Peninsula, Alaska: Polar Research, v. 28, no. 1, p. 71–88, doi:10.1111/j.1751-8369.2009.00102.x.

Raupach, M.R., Marland, G., Ciais, P., Le Quéré, C., Canadell, J.G., Klepper, G., and Field, C.B., 2007, Global and regional drivers of accelerating CO_2 emissions: Proceedings National Academy of Sciences, v. 104, no. 24, p. 10, p. 288–93.

Raven, J., Caldeira, K., Elderfield, H., Hoegh-Guldberg, O., Liss, P., Riebesell, U., Shepherd, J., Turley, C., Waston, A., Heap, R., Banes, R., and Quinn R., 2005, Ocean acidification due to increasing atmospheric carbon dioxide: London, The Royal Society.

Ravens, T., Jones, B.M., Zhang, J., Arp, C.D., and Schmutz, J.A., 2012, Process-based coastal erosion modeling for drew point (North Slope, Alaska): Journal of Waterway, Port, Coastal, and Ocean Engineering, v. 138, no. 2, p. 122–130, doi:10.1061/(ASCE)WW.1943-5460.0000106.

Rawlins, M.A., Steele, M., Holland, M.M., Adam, J.C., Cherry, J.E., Francis, J.A., Groisman, P.Y., Hinzman, L.D., Huntington, T.G., Kane, D.L., Kimball, J.S., Kwok, R., Lammers, R.B., Lee, C.M., Lettenmaier, D.P., McDonald, K.C., Podest, E., Pundsack, J.W., Rudels, B., Serreze, M.C., Shiklomanov, A., Skagseth, Ø., Troy, T.J., Vörösmarty, C.J., Wensnahan, M., Wood, E.F., Woodgate, R., Yang, D., Zhang, K., and Zhang, T., 2010, Analysis of the Arctic system for freshwater cycle intensification—Observations and expectations: Journal of Climate, v. 23, no. 21, p. 5,715–5,737, doi:10.1175/2010JCLI3421.1.

Ray, G., Mccormickray, J., Berg, P., and Epstein, H., 2006, Pacific walrus—Benthic bioturbator of Beringia: Journal of Experimental Marine Biology and Ecology, v. 330, p. 403–419.

Raymond, P.A., McClelland, J.W., Holmes, R.M., Zhulidov, A.V., Mull, Peterson, B.J., Striegl, R.G., Aiken, G.R., and Gurtovaya, T.T., 2007, Flux and age of dissolved organic carbon exported to the Arctic Ocean—A carbon isotopic study of the five largest arctic rivers: Global Biogeochemical Cycles, v. 21, GB4011, 9 p., doi:10.1029/2007GB002934.

Regehr, E.V., Hunter, C.M., Caswell, H., Amstrup, S.C., and Stirling, I., 2009, Survival and breeding of polar bears in the southern Beaufort Sea in relation to sea ice: Journal of Animal Ecology, v. 79, no. 1, p. 117–127, doi:10.1111/j.1365-2656.2009.01603.x.

Regehr, E.V., Lunn, N. J., Amstrup, S.C., and Stirling, I., 2007, Effects of earlier sea ice breakup on survival and population size of polar bears in western Hudson Bay: Journal of Wildlife Management, v. 71, no. 8, p. 2,673–2,683.

Reiners, W.A., Worley, I.A., and Lawrence, D.B., 1971, Plant diversity in a chronosequence at Glacier Bay, Alaska: Ecology, v. 52, p. 55–69.

Ridgewell, A., and Zeebe, R.E., 2005, The role of the global carbonate cycle in the regulation and evolution of the Earth system: Earth and Planetary Science Letters, v. 234, no. 3–4, p. 299–315.

Ringer, G., 2010, Beyond the cruise—navigating sustainable policy and practice in Alaska's Inland Passage, in Luck, M., Maher, P.T., and Stewart, E.J., eds., Cruise tourism in polar regions—Promoting environmental and social sustainability?: Washington, D.C., Earthscan, p. 205–224.

Riordan, B., Verbyla, D., and McGuire, A.D., 2006, Shrinking ponds in subarctic Alaska based on 1950–2002 remotely sensed images: Journal of Geophysical Research, v .111, G04002, 11 p., doi:10.1029/2005JG000150.

Roach, J., 2011, Lake area change in Alaskan National Wildlife Refuges - Magnitude, mechanisms, and heterogeneity: Ph.D. Dissertation. University of Alaska Fairbanks, 225p.

Roach, J., Griffith, B., Verbyla, D., and Jones, J., 2011, Mechanisms influencing changes in lake area in Alaskan boreal forest: Global Change Biology, v. 17, no. 8, p. 2,567–2,583.

Robertson, G., and Brooks, D., 2001, Assessment of the competitive position of the forest products sector in southeast Alaska, 1985–94: General Technical Report PNW-GTR-504, USDA Forest Service, Pacific Northwest Research Station, Portland, Oregon, USA: Website, accessed September 12, 2012, at http://treesearch.fs.fed.us/pubs/2926.

Robinson, D.A., 1993, Monitoring northern hemisphere snow cover: Snow watch '92—Decision Strategies for snow and ice, Glaciological Data Report GD-25, p. 1–25.

Rode, K.D., Amstrup, S.C., and Regehr, E.V., 2010, Reduced body size and cub recruitment in polar bears associated with sea ice decline: Ecological Applications, v. 20, p. 768–782.

Rode, K.D., Peacock, E., Taylor, M., Stirling, I., Born, E.W., Laidre, K.L., and Wiig, Ø., 2012, A tale of two polar bear populations—Ice habitat, harvest, and body condition: Population Ecology, v. 54, p. 3–18.

Romanovsky, V., Burgess, M., Smith, S., Yoshikawa, K., and Brown, J., 2002, Permafrost temperature records—Indicators of climate change: Eos Transactions, American Geophysical Union, v. 83, no. 50, p. 589.

Romanovsky, V.E., Drozdov, D.S., Oberman, N.G., Malkova, G.V., Kholodov, A.L., Marchenko, S.S., Moskalenko, N.G., Sergeev, D.O., Ukraintseva, N.G., Abramov, A.A., Gilichinsky, D.A., and Vasiliev, A.A., 2010b, Thermal state of permafrost in Russia: Permafrost and Periglacial Processes, v. 21, p. 136–155.

Romanovsky, V.E., Gruber, S., Instanes, A., Jin, H., Marchenko, S.S., Smith, S.L., Trombotto, D., and Walter, K.M., 2007, Frozen ground, chap. 7 of Global outlook for ice and snow: Arendal, Norway, Earthprint, UNEP/GRID, p. 181–200.

Romanovsky, V., Oberman, N., Drozdov, D., Malkova, G., Kholodov, A., and Marchenko, S., 2011, Permafrost, in State of the Climate in 2010: Bulletin of the American Meteorological Society, v. 92, no. 6, p. S152–S153.

Romanovsky, V.E., and Osterkamp, T.E., 2000, Effects of unfrozen water on heat and mass transport processes in the active layer and permafrost: Permafrost and Periglacial Processes, v. 11, no. 3, p. 219–239, doi:10.1002/1099-1530(200007/09)11:3<219::AID-PPP352>3.0.CO;2-7.

Romanovsky, V.E., and Osterkamp, T.E., 2001, Permafrost—Changes and impacts, in Paepe, R., and Melnikov, V.P., eds., Permafrost response on economic development, environmental security and natural resources: The Netherlands, Kluwer Academic Publishers, p. 297–315.

Romanovsky, V.E., Smith, S.L., and Christiansen, H.H., 2010a, Permafrost thermal state in the polar northern hemisphere during the International Polar Year 2007–2009—A synthesis: Permafrost and Periglacial Processes, v. 21, no. 2., p. 106–116.

Rosen, Y., 2007, A melting Alaska draws visitors: The Christian Science Monitor, published online November 14, 2007, accessed June 11, 2012, at http://www.csmonitor.com/2007/1114/p03s02-usgn.html.

Rover, J., Ji, L., Wylie, B.K., and Tiezen, L.L., 2012, Establishing water body areal extent trends in interior Alaska from multi-temporal Landsat data: Remote Sensing Letters, v. 3, no. 7, p. 595–604, doi:10.1080/01431161.2011.643507.

Rowland, J.C., Jones, C.E., Altmann, G., Bryan, R., Crosby, B.T., Hinzman, L.D., Kane, D.L., Lawrence, D.M., Mancino, A., Marsh, P., McNamara, J.P., Romanovsky, V.E., Toniolo, H., Travis, B.J., Trochim, E., Wilson, C.J., and Geernaert, G.L., 2010, Arctic landscapes in transition—Responses to thawing permafrost: EOS Transitions, American Geophysical Union, v. 91, no. 26, p. 229.

Rowland, J.C., Travis, B.J., and Wilson, C.J., 2011, The role of advective heat transport in talik development beneath lakes and ponds in discontinuous permafrost: Geophysical Research Letters, v. 38, L17504, 5 p., doi:10.1029/2011GL048497.

Royer, T.C., 1982, Coastal fresh water discharge in the northeast Pacific: Journal of Geophysical Research, v. 87, no. C3, p. 2,017–2,021. doi:10.1029/JC087iC03p02017.

Royer, T.C., and Grosch, C.E., 2006, Ocean warming and freshening in the northern Gulf of Alaska: Geophysical Research Letters, v. 33, L16605, doi:10.1029/2006GL026767.

Ruiz, G.M., Fofonoff, P.W., Carlton, J.T., Wonham, M.J., and Hines, A.H., 2000, Invasion of coastal marine communities in North America—Apparent patterns, processes, and biases: Annual Review of Ecology and Systematics, v. 31, p. 481–531.

Ruiz, G.M., and Hewitt, C.L., 2009, Latitudinal patterns of biological invasions in marine ecosystems—A polar perspective, in Krupnik, I., Lang, M.A., and Miller, S.E., eds., Smithsonian at the poles—Contributions to international polar year science: Washington, D.C., Smithsonian Institution Scholarly Press, p. 347–358, accessed June 11, 2012, at http://si-pddr.si.edu/jspui/bitstream/10088/6824/3/26_Ruiz_pg347-358_Poles.pdf.

Ruiz, G.M., Huber, T., Larson, K., McCann, L., Steves, B., Fofonoff, P., and Hines, A.H., 2006, Biological invasions in Alaska's coastal marine ecosystems—Establishing a baseline: Final report submitted to Prince William Sound Regional Citizens' Advisory Council and the U.S. Fish and Wildlife Service, 112 p., accessed June 11, 2012, at http://www.pwsrcac.org/docs/d0032100.pdf.

Rupp, T.S., Olson, M., Adams, L.G., Dale, B.W., Joly, K., Henkelman, J., Collins, W.B., and Starfield, A.M., 2006, Simulating the influences of various fire regimes on caribou winter habitat: Ecological Applications, v. 16, no. 5, p. 1,730–1,743.

Sakai, A.K., Allendorf, F.W., Holt, J.S., Lodge, D.M., Molofsky, J., Baughman, S., Cabin, R.J., Cohen, J.E., Ellstrand, N.C., McCauley, D.E., O'Neil, P., Parker, I.M., Thompson, J.N., and Weller, S.G., 2001, The population biology of invasive species: Annual Review of Ecology and Systematics, v. 32, p. 305–332.

Salathé, E.P., Jr., 2006, Influences of a shift in North Pacific storm tracks on western North American precipitation under global warming: Geophysical Research Letters, v. 33, L19820, 4 p., doi:10.1029/2006GL026882.

Satellite Imaging Corporation, 2012, SPOT-5 Satellite Sensor, website, accessed August 10, 2012, at http://www.satimagingcorp.com/satellite-sensors/spot-5.html.

Schaberg, P.G., D'Amore, D.V., Hennon, P.E., Halman, J.M., and Hawley, G.J., 2011, Do limited cold tolerance and shallow depth of roots contribute to yellow-cedar decline?: Forest Ecology and Management, v. 262, no. 12, p. 2,142–2,150.

Schlichter, S., 2011, Climate change travel—The world's most endangered places: Independent Traveler.com website, accessed June 11, 2012, at http://www.independenttraveler.com/travel-tips/none/climate-change-travel-the-worlds-most-endangered-places.

Schliebe, S., Rode, K.D., Gleason, J.S., Wilder, J., Proffitt, K., Evans, T.J., and Miller, S., 2008, Effects of sea ice extent and food availability on spatial and temporal distribution of polar bears during the fall open-water period in the Southern Beaufort Sea: Polar Biology, v. 31, no. 8, p. 999–1,010, doi:10.1007/s00300-008-0439-7123.

Schroeder, R.F., and Kookesh, M., 1990, Subsistence harvest and use of fish and wildlife resources and the effects of forest management in Hoonah, Alaska: Juneau, Alaska, Alaska Department of Fish and Game, Division of Subsistence, Technical Paper 142, accessed June 11, 2012, at http://www.adfg.alaska.gov/techpap/tp142.pdf.

Schroth, A.W., Crusius, J., Chever, F., Bostick, B.C., and Rouxel, O.J., 2011, Glacial influence on the geochemistry of riverine iron fluxes to the Gulf of Alaska and effects of deglaciation: Geophysical Research Letters, v. 38, L16605, 6 p., doi:10.1029/2011GL048367.

Schulz, B., 2003, Changes in downed and dead woody material following a spruce beetle outbreak on the Kenai Peninsula, Alaska: U.S. Department of Agriculture, U.S. Forest Service, Research Paper PNW-RP-559, 9 p., accessed June 11, 2012, at http://www.fs.fed.us/pnw/pubs/pnw_rp559.pdf.

Schuster, P., Striegl, G., Aiken, D., Krabbenhoft, D., Dewild, J., Bulter, K., Kamark, B., and Dornblaster, M., 2011, Mercury export from the Yukon River Basin and potential response to a changing climate: Environmental Science and Technology, v. 45, no. 21, p. 9,262–9,267.

Schuur, E.A.G., Abbott, B.W., Bowden, W.B., Brovkin, V., Camill, P., Canadell, J.P., Chapin, F.S., III, Christensen, T.R., Chanton, J.P., Ciais, P., Crill, P.M., Crosby, B.T., Czimczik, C.I., Grosse, G., Hayes, D.J., Hugelius, G., Jastrow, J.D., Kleinen, T., Koven, C.D., Krinner, G., Kuhry, P., Lawrence, D.M., Natali, S.M., Ping, C.L., Rinke, A., Riley, W.J., Romanovsky, V.E., Sannel, A.B.K., Schädel, C., Schaefer, K., Subin, Z.M., Tarnocai, C., Turetsky, M., Walter-Anthony, K. M., Wilson, C.J., and Zimov, S.A., 2011, High risk of permafrost thaw: Nature, v. 480, p. 32–33, doi:10.1038/480032a.

Schuur, E.A.G., Vogel, J.G., Crummer, K.G., Lee, H., Sickman, J.O., and Osterkamp, T.E., 2009, The effect of permafrost thaw on old carbon release and net carbon exchange from tundra: Nature, v. 459, no. 7246, p. 556–559.

Searcy, C., Dean, K., and Stringer, W., 1996, A river–coastal sea ice interaction model—Mackenzie River Delta: Journal of Geophysical Research, v. 101, p. 8,885–8,894.

Selkowitz, D.J., and Stehman, S.V., 2011, Thematic accuracy of the National Land Cover Database (NLCD) 2001 land cover for Alaska: Remote Sensing of Environment, v. 115, no. 6, p. 1,401–1,407.

Serreze, M.C., Walsh, J.E., Chapin, F.S., III, Osterkamp, T., Dyurgerov, M., Romanovsky, V., Oechel, W.C., Morison, J., Zhang, T., and Barry, R.G., 2000, Observational evidence of recent change in the northern high latitude environment: Climatic Change, v. 46, no. 1–2, p. 159–207.

Shasby, M., and Carneggie, D., 1986, Vegetation and terrain mapping in Alaska using Landsat MSS and digital terrain data: Photogrammetric Engineering and Remote Sensing, v. 56, no. 6, p. 779–786.

Shell, 2012, Shell in Alaska, website, accessed August 10, 2012, at http://www.shell.us/home/content/usa/aboutshell/projects_locations/alaska/.

Shulski, M., and Wendler, G., 2007, The climate of Alaska: Fairbanks, Alaska, University of Alaska Press, 214 p.

Shur, Y.L., and Jorgenson, M.T., 2007, Patterns of permafrost formation and degradation in relation to climate and ecosystems: Permafrost and Periglacial Processes, v. 18, no. 1, p. 7–19.

Shvidenko, A., and Apps, M., 2006, The international boreal forest research association—Understanding boreal forests and forestry in a changing world: Mitigation and Adaptation Strategies for Global Change, v. 11, no. 1, p. 5–32.

Smith, L.C., Sheng, Y., MacDonald, G.M., and Hinzman, L.D., 2005, Disappearing arctic lakes: Science, v. 308, p. 1,429–1,429, doi:10.1126/science.1108142.

Smith, S.L., Romanovsky, V.E., Lewkowicz, A.G., Burn, C.R., Allard, M., Clow, G.D., Yoshikawa, K., and Throop, J., 2010, Thermal state of permafrost in North America—A contribution to the international polar year: Permafrost and Periglacial Processes, v. 21, p. 117–135.

Smithsonian Environmental Research Center, 2010, New Introduction—Invasive tunicate found in Whiting Harbor, Sitka, AK: Smithsonian Environmental Research Center, Marine Invasions Research Lab Feature Story, December 2010, accessed June 11, 2012, at http://www.serc.si.edu/labs/marine_invasions/feature_story/December_2010.aspx.

Smol, J.P., and Douglas, M.S.V., 2007, Crossing the final ecological threshold in high Arctic ponds: Proceedings of the National Academy of Sciences, v. 104, no. 30, p. 12,395–12,397, doi:10.1073/pnas.0702777104.

Solomon, S., Qin, D., Manning, M., Chen, Z., Marquis, M., Averyt, K.B., Tignor, M., and Miller, H.L., eds., 2007, Climate change 2007—The physical science basis—Contribution of Working Group I to the Fourth Assessment Report of the Intergovernmental Panel on Climate Change: Cambridge, United Kingdom, and New York, Cambridge University Press, 996 p., accessed June 8, 2012, at http://www.ipcc.ch/publications_and_data/ar4/wg1/en/contents.html.

Spencer, P.D., 2008, Density-independent and density-dependent factors affecting temporal changes in spatial distributions of eastern Bering Sea flatfish: Fisheries Oceanography, v.17, p.396 - 410.

Spies, R.B., and Weingartner, T.J., 2007, Chapter 4 - Long-term change In Spies, R.B., editor, Long-Term Ecological Change in the Northern Gulf of Alaska, Elsevier, Amsterdam, p 259-418, ISBN 9780444529602, 10.1016/B978-044452960-2/50005-9, accessed July 30, 2012 at http://www.sciencedirect.com/science/article/pii/B9780444529602500059.

Stabeno, P.J., Bond, N.A., and Salo, S.A, 2007, On the recent warming of the southeastern Bering Sea shelf: Deep Sea Research, Part II, v. 54, no. 23–26, p. 2,599–2,618.

Stabeno, P.J., Farley, E.V., Jr., Kachel, N.B., Moore, Sue, Mordy, C.W., Napp, J.M., Overland, J.E., Pinchuk, A.I., and Sigler, M.F., 2012, A comparison of the physics of the northern and southern shelves of the eastern Bering Sea and some implications for the ecosystem: Deep Sea Research, Part II, v. 65-70, p. 14–30.

Stabeno, P.J., Napp, J.M., Mordy, C.W., and Whitledge, T.E., 2010, Factors influencing physical structure and lower trophic levels of the eastern Bering Sea shelf in 2005—Sea ice, tides and winds: Progress in Oceanography, v. 85, p. 180–196.

Stabeno, P. J., Schumacher, J.D., and Ohtani, K., 1999, The physical oceanography of the Bering Sea, in Loughlin, T.R., and Ohtani, K., eds., Dynamics of the Bering Sea—A summary of physical, chemical, and biological characteristics, and a synopsis of research on the Bering Sea: Fairbanks, Alaska, North Pacific Marine Science Organization (PICES), University of Alaska Sea Grant, AK-SG-99-03, p. 1–28, accessed June 11, 2012, at http://www.pmel.noaa.gov/pubs/outstand/stab1878/stab1878.shtml.

Stachowicz, J.J., Terwin, H.H., Whitlatch, R.B., and Osman, R.W., 2002, Linking climate change and biological invasions—Ocean warming facilitates nonindigenous species invasions: Proceedings of the National Academy of Sciences, v. 99, no. 24, p. 15,497–15,500.

Stafford, J., Wendler, G., and Curtis, J., 2000, Temperature and precipitation of Alaska—50 Year Trend Analysis: Theoretical and Applied Climatology, v. 67, p. 33–44.

State of Alaska, 2008, Immediate Action Workgroup Recommendations Report to the Governor's Subcabinet on Climate Change, 168 p., accessed June 11, 2012, at http://www.climatechange.alaska.gov/docs/iaw_finalrpt_12mar09.pdf.

State of Alaska, 2010, Alaska's climate change strategy—Addressing impacts in Alaska: Final report submitted by the Adaptation Advisory Group to the Alaska Climate Change Sub-Cabinet, accessed June 11, 2012, at http://www.climatechange.alaska.gov/aag/docs/aag_all_rpt_27jan10.pdf.

State of Alaska, 2011, Climate Change in Alaska, website, accessed August 10, 2012, at http://www.climatechange.alaska.gov/.

Statoil, 2012, Chukchi Sea, Alaska, website, accessed August 10, 2012, at http://www.statoil.com/en/About/Worldwide/NorthAmerica/USA/Alaska/Pages/default.aspx.

Steele, M.A., Ermold, W., and Zhang, J., 2008, Arctic Ocean surface warming trends over the past 100 years: Geophysical Research Letters, v.35, L02614, 6p., doi:10.1029/2007GL031651

Steinacher, M., Joos, F., Frolicher, T.L., Platter, G.-K., and Doney, S.C., 2009, Imminent ocean acidification of the Arctic projected with the NCAR global coupled carbon-cycle climate model: Biogeosciences, v. 6, no. 4, p. 515–533.

Stevenson, D.E., and Lauth, R.R., 2012, Latitudinal trends and temporal shifts in the catch composition of bottom trawls conducted on the eastern Bering Sea shelf: Deep Sea Research, Part II, v. 65-70, p. 251–259, 10.1016/j.dsr2.2012.02.021.

Stewart, B.C., 2011, Changes in frequency of extreme temperature and precipitation events in Alaska: Master's Thesis, Department of Atmospheric Sciences, University of Illinois at Urbana-Champaign.

Stewart, E.J., Howell, S.E.L., Draper, D., Yackel, J., and Tivy, A., 2007, Sea ice in Canada's Arctic—Implication for cruise tourism: Arctic, v. 60, no. 4, p. 370–80.

Stirling, I., Lunn, M.J., and Iacozza, J., 1999, Long-term trends in the population ecology of polar bears in western Hudson Bay in relation to climate change: Arctic, v. 52, p. 294–306.

Stirling, I., and Parkinson, C.L., 2006, Possible effects of climate warming on selected populations of polar bears (Ursus maritimus) in the Canadian Arctic: Arctic, v. 59, no. 3, p. 261–275.

Stone, R.S., Dutton, E.G., Harris, J.M., and Longenecker, D., 2002, Earlier spring snowmelt in northern Alaska as an indicator of climate change: Journal of Geophysical Research, v. 107, no. 4089, 13 p., doi:10.1029/2000JD000286.

Stram, D.L., and Evans, D.C. K., 2009, Fishery management responses to climate change in the North Pacific: International Council for the Exploration of the Sea (ICES), Journal of Marine Science, v. 66, p. 1,633–1,639, doi:10.1093/icesjms/fsp138.

Strategic Environmental Research and Development Program, 2012, Resource Conservation and Climate Change: Website, accessed June 11, 2012, at http://serdp.org/Program-Areas/Resource-Conservation-and-Climate-Change.

Stroeve, J., Holland, M., Meier, W., Scambos, T., and Serreze, M., 2007, Arctic sea ice decline—Faster than forecast: Geophysical Research Letters, v. 34, L09501, 5 p., doi:10.1029/2007GL029703.

Stroeve, J.C., Serreze, M.C., Holland, M.M., Kay, J.E., Malanik, J., and Barrett, A.P., 2011, The Arctic's rapidly shrinking sea ice cover—A research synthesis: Climatic Change, v. 110, no. 3-4, p. 1,005-1,027, doi:10.1007/s10584-011-0101-1.

Sturm, M., Racine, C., and Tape, K., 2001, Increasing shrub abundance in the Arctic: Nature, v. 411, p. 546–547.

Szumigala, D.J., Harbo, L.A., and Alderman, J.N., 2011, Alaska's mineral industry 2010: Alaska Division of Geological and Geophysical Surveys Special Report 65, 83 p., accessed June 11, 2012, at http://www.dggs.dnr.state. ak.us/pubs/id/22822.

Talbot, S.S., and Markon, C.J., 1986, Vegetation mapping of Nowitna National Wildlife Refuge, Alaska using Landsat MSS digital data: Photogrammetric Engineering and Remote Sensing, v. 52, no. 6, p. 791–799.

Talbot, S.S., and Markon, C.J., 1988, Intermediate-scale vegetation mapping of Innoko National Wildlife Refuge, Alaska using Landsat MSS digital data: Photogrammetric Engineering and Remote Sensing, v. 54, no. 3, p. 377–383.

Tan, Z., Tieszen, L.L., Zhu, Z., Liu, S., and Howard, S.M., 2007, An estimate of carbon emissions from 200 wildfires across Alaskan Yukon River basin: Carbon Balance and Management, v. 2, no. 12, 8 p., doi:10.1186/1750-0680-2-12.

Tape, K., Sturm, M., and Racine, C., 2006, The evidence for shrub expansion in northern Alaska and Pan-Arctic: Global Change Biology, v. 12, no. 4, p. 686–702.

Tape, K., Verbyla, D., and Welker, J.M., 2011, Twentieth century erosion in Arctic Alaska foothills—The influence of shrubs, runoff, and permafrost: Journal of Geophysical Research—Biogeosciences, v. 116, G04024, 11 p., doi:10.1029/2011JG001795.

Tarnocai, C., Canadell, J.G., Schuur, E.A.G., Kuhry, P., Mazhitova, G., and Zimov, S., 2009, Soil organic carbon pools in the northern circumpolar permafrost region: Global Biogeochemical Cycles, v. 23, GB2023, 11 p., doi:10.1029/2008GB003327.

Trainor, S.F., Calef, M., Natcher, D., Chapin, F.S., III, McGuire, A.D., Huntington, O., Duffy, P., Rupp, T.S., DeWilde, L., Kwart, M., Fresco, N., and Lovecraft, A.L., 2009, Vulnerability and adaptation to climate-related impacts in rural and urban interior Alaska: Polar Research, v. 28, no. 1, p. 100–118.

University of Alaska Anchorage, Alaska Natural Heritage Program, 2012a, what is AKEPIC? website, accessed August 10, 2012, at http://aknhp.uaa.alaska.edu/botany/akepic.

University of Alaska Anchorage, Alaska Natural Heritage Program, 2012b, BLM-Rapid Ecoregoinal Assessments, website, accessed August 10, 2012, at http://aknhp.uaa. alaska.edu/ecology/projects/blm-rapid-ecoregional-assessments/.

University of Alaska Fairbanks, Alaska Center for Climate Assessment and Policy, 2011, Spring breakup summary in Alaska climate dispatch–A state–wide seasonal summary and outlook, Gamble, B., ed.: University of Alaska–Fairbanks, Alaska Center for Climate Assessment and Policy, summer 2011 issue, p. 4–5, accessed June 7, 2012, at http://ine.uaf.edu/accap/documents/summer11_dispatch.pdf.

University of Alaska Fairbanks, 2012a, SNAP, Scenarios Network for Alaska and Arctic Planning: website, accessed August 10, 2012, at http://www.snap.uaf.edu/.

University of Alaska Fairbanks, 2012b, Alaska Center for climate Assessment and Policy, website, accessed August 10, 2012, at http://www.accap.uaf.edu.

University of Alaska Fairbanks, 2012c, Project Jukebox, website, accessed August 10, 2012, at http://jukebox.uaf. edu/site/.

University of Alaska Fairbanks, 2012d, Alaska Center for Climate Assessment and Policy – Alaska Climate Dispatch, website, accessed August 10, 2012, at http://ine.uaf.edu/accap/dispatch.htm.

University of Alaska Fairbanks, 2012e, Alaska Weather and Climate Highlights, website, accessed August 10, 2012, at http://ine.uaf.edu/accap/awch/about.htm.

University of Alaska Fairbanks, 2012f, Permafrost Laboratory, website, accessed August 10, 2012, at http://permafrost. gi.alaska.edu/.

University of Alaska Fairbanks, 2012g, Ocean Acidification Research Center, website, accessed October 2, 2012, at http://www.sfos.uaf.edu/oarc/index.php.

U.S. Arctic Research Commission, 2010, The role of scaling in societal applications—public and private infrastructure vulnerabilities, in Vorosmarty, C., McGuire, D., and Hobbie, J., eds., Scaling studies in Arctic system science and policy support: Arlington, Va., U.S. Arctic Research Commission, accessed June 11, 2012, at http://www.arctic. gov/publications/arctic_scaling.pdf.

U.S. Arctic Research Commission Permafrost Task Force, 2003, Climate change, permafrost, and impacts on civil infrastructure: Arlington, Va., U.S. Arctic Research Commission, Special Report 01–03, accessed June 11, 2012, at http://www.arctic.gov/publications/permafrost.pdf.

U.S. Army Corps of Engineers, 2006, An examination of erosion issues in the communities of Bethel, Dillingham, Kaktovik, Kivalina, Newtok, Shishmaref, and Unalakleet: Alaska Village Erosion Technical Assistance Program, accessed June 11, 2012, at http://www.housemajority.org/coms/cli/AVETA_Report.pdf.

U.S. Bureau of Ocean Energy Management, 2012, Application of high frequency radar to potential hydrocarbon development areas in the Northeast Chukchi Sea: BOEM Environmental Studies Program—Ongoing Studies, accessed October 2, 2012, at http://www.boem.gov/uploadedFiles/BOEM/Environmental_Stewardship/Environmental_Studies/Alaska_Region/Alaska_Studies/PO_0906.pdf.

U.S. Census Bureau, 2010, Alaska: Website, accessed June 11, 2012, at http://quickfacts.census.gov/qfd/states/02000.html.

U.S. Department of Agriculture, 2009, 2007 Census of Agriculture, Alaska State and Area, Vol. 1: U.S. Department of Agriculture Geographic Area Series Part 1, AC-07-A-2, 373 p., accessed June 11, 2012, at http://www.agcensus.usda.gov/Publications/2007/Full_Report/index.asp.

U.S. Department of Agriculture, Forest Service, 2001, Forest health protection report—Forest insect and disease conditions in Alaska—2000: U.S. Department of Agriculture, U.S. Forest Service, General Technical Report R10-TP-86, 62 p., accessed June 11, 2011, at http://forestry.alaska.gov/pdfs/00part1.pdf.

U.S. Department of Agriculture, Forest Service, 2002, Revised land and resource management plan—Chugach National Forest: U.S. Department of Agriculture, Forest Service, Alaska Region Chugach National Forest RIO-MB-480C, 329 p., accessed September 11, 2012, at http://www.fs.usda.gov/Internet/FSE_DOCUMENTS/fsm8_028736.pdf.

U.S. Department of Defense, 2010, Quadrennial defense review report: U.S. Department of Defense, accessed June 11, 2012, at http://www.defense.gov/qdr/images/QDR_as_of_12Feb10_1000.pdf.

U.S. Department of Defense, 2011, Report to Congress on Arctic operations and the Northwest Passage, 32 p., accessed June 11, 2012, at http://www.defense.gov/pubs/pdfs/Tab_A_Arctic_Report_Public.pdf.

U.S. Department of the Interior, Bureau of Land Management, 2012, National Petroleum Reserve–Alaska Draft Integrated Activity Plan–Environmental Impact Statement: U.S. Department of the Interior, Bureau of Land Management.

U.S. Environmental Protection Agency, 2011, Climate Change Impacts and Adapting to Change: U.S. Environmental Protection Agency Climate Change—Impacts and Adaptation website page, accessed July 24, 2012, at http://www.epa.gov/climatechange/impacts-adaptation/

U.S. General Accounting Office, 2003, Alaska Native villages—Most are affected by flooding and erosion, but few qualify for Federal assistance: U.S. General Accounting Office Report GAO-04-142, 85 p., accessed June 11, 2012, at http://www.gao.gov/new.items/d04142.pdf.

U.S. Geological Survey, 2006, Assessing the impacts of climate variability and change on the Nation's resources— Impacts on Interior resources: U.S. Geological Survey, accessed June 11, 2012, at http://esp.cr.usgs.gov/info/assessment/.

U.S. Geological Survey, 2012a, Real-Time Permafrost and Climate Monitoring Network, Arctic Alaska, website, accessed August 10, 2012, at http://data.usgs.gov/climateMonitoring/region/show?region=alaska.

U.S. Geological Survey, 2012b, Cold Regions Lake and Landscape Research, website, accessed August 10, 2012, at http://alaska.usgs.gov/science/geography/studies/index.php.

U.S. Geological Survey, 2012c, Land Processes Distributed Active Archive Center–MODIS Products Table, website, accessed August 10, 2012, at https://lpdaac.usgs.gov/products/modis_products_table.

U.S. Geological Survey, 2012d, Landsat Missions, website, accessed August 10, 2012, at http://landsat.usgs.gov/index.php.

U.S. Government Accountability Office, 2011, Coast Guard observations on Arctic requirements, icebreakers, and coordination with stakeholders: U.S. Government Accountability Office Report GAO-12-254T, accessed June 11, 2012, at http://www.gao.gov/products/GAO-12-254T.

VanLooy, J., Forster, R., and Ford, A., 2006, Accelerating thinning of Kenai Peninsula glaciers, Alaska: Geophysical Research Letters, v. 33, L21307, 5 p., doi:10.1029/2006GL028060.

Verbyla, D., 2008, The greening and browning of Alaska based on 1982–2003 satellite data: Global Ecology and Biogeography, v. 17, no. 4, p. 547–555.

Vorosmarty, C., Hinzman, L., and Pundsack, J., 2008, Introduction to special section on changes in the Arctic freshwater system—Identification, attribution, and impacts at local and global scales: Journal of Geophysical Research, v. 113, G01S91, 5 p., doi:10.1029/2007JG000615.

Wadley, M.R., and Bigg, G.R., 2002, Impact of flow through the Canadian Archipelago and Bering Strait on the North Atlantic and Arctic circulation—An ocean modeling study: Quarterly Journal of the Royal Meteorological Society, v. 128, no. 585, p. 2,187–2,203.

Walsh, J.E., Chapman, W.L., Romanovsky, V., Christensen, J.H., and Stendel, M., 2008, Global climate model performance over Alaska and Greenland: Journal of Climate, v. 21, p. S6156–S6174.

Walsh, J.J., McRoy, C.P., Coachman, L.K., Goering, J.J., Nihoul, J.J., Whirledge, T.E., Blackburn, T.H., Parker, P.L., Wirick, C.D., Shuert, P.G., Grebmeier, J.M., Springer, A.W., Tripp, R.D., Hansell, D.A., Djenidi, S., Deleersnijder, E., Henriksen, K., Lund, B.A., Andersen, P., Muller-Karger, F.E., and Dean, K., 1989, Carbon and nitrogen cycling within the Bering/Chukchi Seas—Source regions for organic matter effecting AOU demands of the Arctic Ocean: Progress in Oceanography, v. 22, no. 4, p. 277–259.

Walter, K.M., Smith, L.C., and Chapin, F.S., III, 2007, Methane bubbling from northern lakes—Present and future contributions to the global methane budget: Philosophical Transactions of the Royal Society A, v. 365, no. 1856, p. 1,657–1,676.

Walvoord, M.A., and Striegl, R.G., 2007, Increased groundwater to stream discharge from permafrost thawing in the Yukon River basin—Potential impacts on lateral export of carbon and nitrogen: Geophysical Research Letters, v. 34, L12402, 6 p., doi:10.1029/2007GL030216.

Wang, M., and Overland, J.E., 2009, A sea ice free summer Arctic within 30 years?: Geophysical Research Letters, v. 36, L07502, 5 p., doi:10.1029/2009GL037820.

Ware, D.M., and Thomson, R.E., 2005, Bottom-up ecosystem trophic dynamics determine fish production in the northeast Pacific: Science, v. 308, p. 1,280–1,284.

Warren, J., Berner, J., and Curtis, T., 2005, Climate change and human health—Infrastructure impacts to small remote communities in the North: International Journal of Circumpolar Health, v. 64, no. 5, p. 487–497.

Weingartner, T.J., Danielson, S.L., and Royer, T.C., 2005, Freshwater variability and predictability in the Alaska coastal current: Deep Sea Research, Part II, v. 52, no. 24–26, p. 169–191.

Weingartner, T.J., Okkonen, S.R., and Danielson, S.L., 2009, Circulation and water property variations in the nearshore Alaskan Beaufort Sea (1999–2007): University of Alaska–Fairbanks, Institute of Marine Science, Final Report, OCS Study MMS 2009-035, 155 p.

Welch, D.W., Ishida, Y., and Nagasawa, K., 1998, Thermal limits and ocean migrations of sockeye salmon (*Oncorhynchus nerka*)—Long–term consequences of global warming: Canadian Journal of Fisheries and Aquatic Science, v. 55, no. 4, p. 937–948.

Weller, G., 2005, Summary and synthesis of the ACIA *in* Arctic Climate Impact Assessment (ACIA): Cambridge, United Kingdom, Cambridge University Press, accessed June 12, 2012, at http://www.acia.uaf.edu.

Wendler, G., and Shulski, M., 2009, A century of climate change for Fairbanks, AK: Arctic, v. 62, no. 3, p. 295–300.

Wendler, G., Shulski, M., and Moore, B., 2010, Changes in the climate of the Alaskan North Slope and the ice concentration of the adjacent Beaufort Sea: Theoretical and Applied Climatology, v. 99, no. 1–2, p. 67–74, doi:10.1007/s00704-009-0127-8.

Western Alaska Community Development Association, 2008, Western Alaska Community Development Quota Program: Western Alaska Community Development Association, 18 p., accessed June 12, 2012, at http://www.wacda.org/media/pdf/SMR_2008.pdf.

Western Regional Climate Center Desert Research Institute, 2011, Climate of Alaska: Western Regional Climate Center Desert Research Institute website, accessed June 12, 2012, at http://www.wrcc.dri.edu/narratives/ALASKA.htm.

White, D.M., Gerlach, S.C., Loring, P.A., and Tidwell, A., 2007, Food and water security in a changing arctic climate: Environmental Research Letters, v. 2, no. 4, 4 p.

Wijffels, S.E., Schmitt, R.W., Bryden, H.L., and Stigebrandt, A., 1992, Transport of freshwater by the oceans: Journal of Physical Oceanography, v. 22, no. 2, p. 155–162.

Wiken, E.B., 1986, Terrestrial ecozones of Canada: Lands Directorate, Environment Canada Ecological Land Classification Series 19, 26 p.

Wilson, W.J., and Ormseth, O.A., 2009, A new management plan for Arctic waters of the United States: Fisheries, v. 34, p. 555–558.

Witherell, D., 2009, Developing ecosystem-based management of U.S. fisheries: Current—Journal of Marine Education, v. 25, no. 3, p. 13–16.

Wolfe, R.J., 2004, Local traditions and subsistence—A synopsis of twenty-five years of research by the State of Alaska: Juneau, Alaska, Alaska Department of Fish and Game, Division of Subsistence Technical Paper 284, 89 p., accessed June 12, 2012, at http://www.subsistence.adfg.state.ak.us/TechPap/tp284Twentyfiveyears.pdf.

Wolken, J.M., Hollingsworth, T.N., Rupp, T.S., Chapin, F.S., III, Trainor, S.F., Barrett, T.M., Sullivan, P.F., McGuire, A.D., Euskirchen, E.S., Hennon, P.E., Beever, E.A., Conn, J.S., Crone, L.S., D'Amore, D.V., Fresco, N., Hanley, T.A., Kielland, K., Kruse, J.J., Patterson, T., Schuur, E.A.G., Verbyla, D.L., and Yarie, J., 2011, Evidence and implications of recent and projected climate change in Alaska's forest ecosystems: Ecosphere, v. 2, no. 11, p. 124.

Woo, M.K., 1992, Impacts of climatic variability and change on Canadian wetlands: Canadian Water Resources Journal, v. 17, p. 63−69.

Woodgate, R.A., Aagaard, K., and Weingartner, T.J., 2005, A year in the physical oceanography of the Chukchi Sea—Moored measurements from autumn 1990−91: Deep Sea Research Part II, v. 52, no. 24−26, p. 3,116−3,149, doi: 10.1016/j.dsr2.2005.10.016.

Woodgate, R.A., Aagaard, K., and Weingartner, T.J., 2006, Interannual changes in the Bering Strait fluxes of volume, heat and freshwater between 1991 and 2004: Geophysical Research Letters, v. 33, L15609, 5 p., doi:10.1029/2006GL026931.

Woodgate, R.A., Weingartner, T., and Lindsay, R., 2010, The 2007 Bering Strait oceanic heat flux and possible relationships to anomalous Arctic Sea-ice retreat: Geophysical Research Letters, v. 37, L01602, 5 p., doi:10.1029/2009GL041621.

Woodson, D., Dello, K., Flint, L., Hamilton, R., Neilson, R., and Winton, J., 2011, Climate change effects in the Klamath basin, chap. 6 of Thorsteinson, L., Vanderkooi, S., and Duffy, W., eds., Proceedings of the Klamath Basin Science Conference, Medford, Oregon, February 1–5, 2010: U.S. Geological Survey Open-File Report 2011–1196, p. 123–150. (Also available at http://pubs.er.usgs.gov/publication/ofr20111196.)

Wylie, B.K., Zhang, L., Ji, L., Tieszen, L.L., and Bliss, N.B., 2008, Modeling and dynamic monitoring of ecosystem performance in the Yukon River Basin: U.S. Geological Survey Fact Sheet 2008–3016, 2 p. (Also available at http://pubs.usgs.gov/fs/2008/3016/.)

Yamamoto-Kawai, M., McLaughlin, F.A., Carmack, E.C., Nishino, S., and Shimada, K., 2009, Aragonite undersaturation in the Arctic Ocean—Effects of ocean acidification and sea ice melt: Science, v. 326, no. 5956, p. 1,098–1,100.

Yarie, J., and Van Cleve, K., 2006, Controls over forest production in Interior Alaska, in Chapin, F.S., III, ed., Alaska's changing boreal forest: United Kingdom, Oxford University Press, p. 171–188.

Yatsu, A., Aydin, Y., King, J.R., McFarlane, G.A., Chiba, S., Tadokoro, K., and Kaeriyama, M., 2008, Elucidating dynamic responses of North Pacific fish populations to climatic forcing influence of life-history strategy: Progress in Oceanography, v. 77, no. 2–3, p. 252–268.

Yin, J.H., 2005, A consistent poleward shift of the storm tracks in simulations of 21st century climate: Geophysical Research Letters, v. 32, L18701, 4 p., doi:10.1029/2005GL023684.

Yoo, J.C., and D'Odorico, P., 2002, Trends and fluctuations in the dates of ice break-up of lakes and rivers in Northern Europe—The effect of the North Atlantic Oscillation: Journal of Hydrology, v. 268, p. 100−112.

Yoshikawa, K., and Hinzman, L.D., 2003, Shrinking thermokarst ponds and groundwater dynamics in discontinuous permafrost: Permafrost and Periglacial Processes, v. 14, no. 2, p. 151−160.

Yu, G., Schwartz, Z., and Walsh, J.E., 2009, A weather-resolving index for assessing the impact of climate change on tourism related climate resources: Climatic Change, v. 95, p. 551–573, doi:10.1007/s10584-009-9565-7.

Zhang, X., 2010, Sensitivity of Arctic summer sea ice coverage to global warming forcing—Towards reducing uncertainty in Arctic climate change projections: Tellus−Series A, v. 62, no. 3, p. 220−227.

Zhuang, Q., Melillo, J.M., McGuire, A.D., Kicklighter, D.W., Prinn, R.G., Steudler, P.A., Felzer, B.S., and Hu, S., 2007, Net emissions of CH_4 and CO_2 in Alaska—Implications for the region's greenhouse gas budget: Ecological Applications, v. 17, no. 1, p. 203−212.

Appendix A. 2012 NCA Alaska Technical Regional Report Writing Team

Carl Markon – Team Lead
Office of the Regional Executive – Alaska
U.S. Geological Survey
markon@usgs.gov

F. Stuart Chapin III
National Climate Assessment–National Climate
Assessment and Development Advisory
Committee, Federal Advisory Committee Act member
Institute of Arctic Biology
University of Alaska Fairbanks
Terry.chapin@alaska.edu

Sarah F. Trainor
Alaska Center for Climate Assessment and Policy
University of Alaska Fairbanks
Sarah.trainor@alaska.edu

Vanessa Skean
Office of the Regional Executive – Alaska
U.S. Geological Survey
vskean@usgs.gov

Stephen Gray
Alaska Climate Science Center
U.S. Geological Survey
sgray@usgs.gov

Michael Brubaker
Center for Climate and Health
Alaska Native Tribal Health Consortium
mbrubaker@anthc.org

Durelle Smith
Office of the Regional Executive – Alaska
U.S. Geological Survey
dpsmith@usgs.gov

Philip Loring
Alaska Center for Climate Assessment and Policy
University of Alaska Fairbanks
ploring@alaska.edu

Jon Zufelt
Cold Regions Research and Engineering Laboratory
U.S. Army Corps of Engineers
Jon.e.zufelt@usace.army.mil

Molly McCammon
Alaska Ocean Observing System
mccammon@aoos.org

James Partain
Alaska Region
National Oceanic and Atmospheric Administration
james.partain@noaa.gov

This page intentionally left blank.

Appendix B. Author List

Regional Description

Alisa Gallant
Earth Resources Observation and Science Center
U.S. Geological Survey
gallant@usgs.gov

Guido Grosse
Geophysical Institute
University of Alaska Fairbanks
ggrosse@gi.alaska.edu

Kevin Hillmer-Pegram
Resilience and Adaptation Program
University of Alaska Fairbanks
khillmerpegram@alaska.edu

Benjamin Jones
Alaska Science Center
U.S. Geological Survey
bjones@usgs.gov

Philip Loring
Alaska Center for Climate Assessment and Policy
University of Alaska Fairbanks
ploring@alaska.edu

Vladimir Romanovsky
Department of Geology and Geophysics
University of Alaska Fairbanks
ffver@uaf.edu

Jane Wolken
School of Natural Resources & Agriculture Science
University of Alaska Fairbanks
jmwolken@alaska.edu

Alaska's Climate Trends

David Atkinson
International Arctic Research Center
University of Victoria, Canada
datkinson@uvic.ca

Eugenie Euskirchen
Institute of Arctic Biology
University of Alaska Fairbanks
seeuskirchen@alaska.edu

Oceana Francis
International Arctic Research Center
University of Alaska Fairbanks
oceana@iarc.uaf.edu

Stephen Gray
Alaska Climate Science Center
U.S. Geological Survey
sgray@usgs.gov

Gleb Panteleev
International Arctic Research Center
University of Alaska Fairbanks
gleb@iarc.uaf.edu

Brooke Stewart
National Climatic Data Center
National Oceanic and Atmospheric Administration
brooke.stewart@noaa.gov

John Walsh
International Arctic Research Center
University of Alaska Fairbanks
jwalsh@iarc.uaf.edu

Regional Climate Forecasts

Kenneth Kunkel
Division of Atmospheric Sciences
Desert Research Institute
Kenneth.Kunkel@dri.edu

Laura Stevens
National Climatic Data Center
National Oceanic and Atmospheric Administration
Laura.Stevens@noaa.gov

Brooke Stewart
National Climatic Data Center
National Oceanic and Atmospheric Administration
brooke.stewart@noaa.gov

John Walsh
International Arctic Research Center
University of Alaska Fairbanks
jwalsh@iarc.uaf.ed

Observed Environmental Trends

David Atkinson
International Arctic Research Center
University of Victoria, Canada
datkinson@uvic.ca

James Berner, M.D.
Center for Climate and Health
Alaska Native Tribal Health Consortium
jberner@anthc.org

Tim Brabets
Alaska Science Center
U.S. Geological Survey
tbrabets@usgs.gov

Todd Brinkman
Scenarios Network for Alaska & Arctic Planning
University of Alaska Fairbanks
tjbrinkman@alaska.edu

Michael Brubaker
Center for Climate and Health
Alaska Native Tribal Health Consortium
mbrubaker@anthc.org

Gary Clow
Geology and Environmental Change Science Center
U.S. Geological Survey
clow@usgs.gov

Observed Environmental Trends—Continued

John Crusius
Woods Hole Field Center
U.S. Geological Survey
jcrusius@usgs.gov

Seth Danielson
Institute of Marine Science
University of Alaska Fairbanks
sldanielson@alaska.edu

Anthony DeGange
Alaska Science Center
U.S. Geological Survey
tdegange@usgs.gov

Hajo Eicken
International Arctic Research Center
University of Alaska Fairbanks
Hajo.eicken@gi.alaska.edu

Eugenie Euskirchen
Institute of Arctic Biology
University of Alaska Fairbanks
seeuskirchen@alaska.edu

Jeff Freymeuller
Geophysical Institute
University of Alaska Fairbanks
Jeff.freymueller@gi.alaska.edu

Stephen Gray
Alaska Climate Science Center
U.S. Geological Survey
sgray@usgs.gov

Jacqueline M Grebmeier
Chesapeake Biological Lab
University of Maryland Center for
Environmental Science
jgrebmeri@umces.edu

Brad Griffith
Alaska Cooperative Fish and Wildlife Research Unit
U.S. Geological Survey
University of Alaska Fairbanks
ffdbg@aurora.alaska.edu

Guido Grosse
Geophysical Institute
University of Alaska Fairbanks
ggrosse@gi.alaska.edu

Observed Environmental Trends—Continued

Victoria Hykes Steere
Alaska Pacific University
vhsteere@alaskapacific.edu

Larry Hinzman
International Arctic Research Center
University of Alaska Fairbanks
lhinzman@iarc.uaf.edu

Teresa Hollingsworth
Pacific Northwest Research Station
U.S.D.A. Forest Service
University of Alaska Fairbanks
tnhollingsworth@alaska.edu

Eran Hood
Department of Natural Science
University of Alaska Southeast
Eran.hood@uas.alaska.edu

Yongwon Kim
International Arctic Research Center
University of Alaska Fairbanks
kimyw@iarc.uaf.edu

Michael R Lilly
GW Scientific
mlilly@gwscientific.com

Scott Lindsey
Alaska-Pacific River Forecast Center
NOAA/National Weather Service
scott.lindsey@noaa.gov

Philip Loring
Alaska Center for Climate Assessment and Policy
University of Alaska Fairbanks
ploring@alaska.edu

Carl Markon
Office of the Regional Executive – Alaska
U.S. Geological Survey
markon@usgs.gov

Molly McCammon
Alaska Ocean Observing System
mccammon@aoos.org

A. David McGuire
U.S. Geological Survey
University of Alaska Fairbanks
admcguire@alaska.edu

Jeremy Mathis
Institute of Marine Science
University of Alaska Fairbanks
jmathis@sfos.uaf.edu

Shad O'Neel
Alaska Science Center
U.S. Geological Survey
soneel@usgs.gov

Thomas M Ravens
School of Engineering
University of Alaska Anchorage
tomravens@uaa.alaska.edu

Lisa L Robbins
St. Petersburg Coastal and Marine Science Center
U.S. Geological Survey
lrobbins@usgs.gov

Tracy Rogers
International Arctic Research Center
University of Alaska Fairbanks
tsrogers@alaska.edu

Vladimir Romanovsky
Department of Geology and Geophysics
University of Alaska Fairbanks
ffver@uaf.edu

Phyllis Stabeno
Pacific Marine Environmental Laboratory
National Oceanic and Atmospheric Administration
phyllis.stabeno@noaa.gov

Martin Truffer
International Arctic Research Center
University of Alaska Fairbanks
truffer@gi.alaska.edu

Thomas Van Pelt
Bering Sea Integrated Ecosystem Research
North Pacific Research Board
tvanpelt@nprb.org

Thomas Weingartner
Institute of Marine Science
University of Alaska Fairbanks
tjweingartner@alaska.edu

Observed Environmental Trends—Continued

Dan White
Institute of Northern Engineering
University of Alaska Fairbanks
dmwhite@alaska.edu

Jane Wolken
School of Natural Resources & Agriculture Science
University of Alaska Fairbanks
jmwolken@alaska.edu

Jon Zufelt
Cold Regions Research and Engineering Laboratory
U.S. Army Corps of Engineers
Jon.e.zufelt@usace.army.mil

Potential Effects from A Changing Climate

Todd Brinkman
Scenarios Network for Alaska & Arctic Planning
University of Alaska Fairbanks
tjbrinkman@alaska.edu

F. Stuart Chapin III
Institute of Arctic Biology
University of Alaska Fairbanks
Terry.chapin@alaska.edu

Gary Clow
Geology and Environmental Change Science Center
U.S. Geological Survey
clow@usgs.gov

Anthony DeGange
Alaska Science Center
U.S. Geological Survey
tdegange@usgs.gov

Stephen Gray
Alaska Climate Science Center
U.S. Geological Survey
sgray@usgs.gov

Guido Grosse
Geophysical Institute
University of Alaska Fairbanks
ggrosse@gi.alaska.edu

Hannah Harrison
Human Dimensions Laboratory
University of Alaska Fairbanks
hharri22@alaska.edu

Potential Effects from A Changing Climate—Continued

Kevin Hillmer-Pegram
Resilience and Adaptation Program
University of Alaska Fairbanks
khillmerpegram@alaska.edu

Teresa Hollingsworth
Pacific Northwest Research Station
USDA Forest Service
University of Alaska Fairbanks
tnhollingsworth@alaska.edu

Enukyoung Hong
Institute of Northern Engineering
University of Alaska Fairbanks
eunkyoung.hong@gmail.com

Dennis Lassuy
Alaska Region
US Fish and Wildlife Service
Currently with the Bureau of Land Management
dlassuy@blm.gov

Philip Loring
Alaska Center for Climate Assessment and Policy
University of Alaska Fairbanks
ploring@alaska.edu

Peter Larsen
Lawrence Berkeley National Laboratory
PHLarsen@lbl.gov

Carl Markon
Office of the Regional Executive – Alaska
U.S. Geological Survey
markon@usgs.gov

James E Overland
Pacific Marine Environmental Laboratory
National Oceanic and Atmospheric Administration
James.E.Overland@noaa.gov

Lyman Thorsteinson
Office of the Regional Executive–Alaska
U.S. Geological Survey
lthorsteinson@usgs.gov

Vladimir Romanovsky
Department of Geology and Geophysics
University of Alaska Fairbanks
ffver@uaf.edu

Potential Effects from A Changing Climate—Continued

Jane Wolken
School of Natural Resources & Agriculture Science
University of Alaska Fairbanks
jmwolken@alaska.edu

New Science Leadership on the Alaskan Landscape

Greg Balogh
Arctic Landscape Conservation Cooperative
U.S. Fish and Wildlife Service
greg_balogh@fws.gov

Michael Brubaker
Center for Climate and Health
Alaska Native Tribal Health Consortium
mbrubaker@anthc.org

Molly McCammon
Alaska Ocean Observing System
mccammon@aoos.org

Brook Gamble
Alaska Center for Climate Assessment & Policy
University of Alaska Fairbanks
Brook.gamble@alaska.edu

Stephen Gray
Alaska Climate Science Center
U.S. Geological Survey
sgray@usgs.gov

Jackie Kramer
Region 10 – Pacific Northwest
U.S. Environmental Protection Agency
Kramer.Jackie@epamail.epa.gov

Carl Markon
Office of the Regional Executive – Alaska
U.S. Geological Survey
markon@usgs.gov

John Payne
North Slope Science Initiative
Bureau of Land Management
jpayne@blm.gov

New Science Leadership on the Alaskan Landscape—Continued

Scott Rupp
Scenarios Network for Alaska and Arctic Planning
University of Alaska Fairbanks
tsrupp@alaska.edu

Sarah F. Trainor
Alaska Center for Climate Assessment and Policy
University of Alaska Fairbanks
Sarah.trainor@alaska.edu

Gerd Wendler
Alaska Climate Research Center
University of Alaska Fairbanks
gerd@dino.gi.alaska.edu

Planning for the Future

F. Stuart Chapin III
Institute of Arctic Biology
University of Alaska Fairbanks
Terry.chapin@alaska.edu

Dave D'Amore
Pacific Northwest Research Station
U.S. Forest Service
ddamore@fs.fed.us

Molly McCammon
Alaska Ocean Observing System
mccammon@aoos.org

Carl Markon
Office of the Regional Executive – Alaska
U.S. Geological Survey
markon@usgs.gov

Sarah F. Trainor
Alaska Center for Climate Assessment and Policy
University of Alaska Fairbanks
sarah.trainor@alaska.edu

Bruce Wylie
EROS Data Center
U.S. Geological Survey
wylie@usgs.gov

This page intentionally left blank.

Appendix C. Alaska Forum on the Environment[1]: Climate Change: Our Voices, Sharing Ways Forward.

Anchorage, Alaska. February 7, 2012, 10:30–11:45 a.m.

Opening Remarks:

Victoria Hykes Steere: Alaska Pacific University.

We ask you to share with us what you've seen, heard, and what you think. Your information will be transcribed and submitted to the Tribal Chapter of the National Climate Assessment.

Wilson Justin: Cheesh'Na Tribal Council.

One of the great obstacles with climate change is that it's predicated on outcomes from data sets. You can't get outcomes from data sets on a 20 year basis. So it's critical to get observations from people on the land. If we let the researchers and data people run loose, we'll be sitting on the sidelines for the next 100 years trying to get a conclusive statement from these researchers.

A vital point that we often overlook in climate change discussions, is that we talk about what has changed, but hardly ever, about what we'll do for adaptation, and there is no one who's going to tell us what to do, to adapt in the next 20 or 230 years. We have a lot of problems to look at on 2 levels—(1) direct observation, and (2) Researchers never talked about with climate change as a sociological issue.

When I first talked about climate change in the 1980's, the prior generation looked at these changes as in "this is God's work, don't criticize the Creator." It appeared as if we were trying to make definitive judgments about climate change, something that the Creator is responsible for, so we lost several decades of indigenous observation, because our people refused to share. Those observations we most needed. In the meantime those in the academic field who had knowledge didn't have the common tools of a language to portray works of Creation scientifically as opposed to criticizing the work of the Creator.

A single observation overheard in the mid 1960s haunted me for years. When I was a teenager up in Slana Alaska, I spent time around elderly prospectors and other old timers. Many were around in early 1900's. One day I happened to be at a place where they gathered. A community well actually, and they were talking about the weather. I heard them say: these leaves are always getting bigger every summer. I thought to myself, "can't they find something more interesting to talk about?" But those words stuck in my mind so I started paying attention to what was growing out there. By 1975 it was clear that there were changes in leaves and trees, but I didn't understand what it meant. By 1985, I began looking at these issues, measuring or marking when spring came and only few short years later realized changes in leaves were due to earlier spring. My family lived at the 3,000 foot level and June was when it started to green up. Then around May 22, the earliest I recorded green was on May 8 in about 1994, why were leaves getting so much bigger? Spring came earlier each decade, by the 1980's I could clearly see this but I had no idea of what they were talking about in 1965. Elderly folks knew changes were happening 20 years before I figured it out. 20 years lead in time before their remarks. Very clear precise observations, but not a single one of us was able to articulate it. Tremendous amount of time went by before we understood what they were talking about.

All your observations accumulated mean something. Science data sets are not going to give conclusions for another generation, we have to do that.

Victoria and I, we work well together. Soft spoken and sweet but she's a dragon lady. She's all bite and I'm all bark and between the two of us we cover it all.

---

[1]Alaska Forum on the Environment, 2012, website, accessed August 10, 2012, at www.akforum.com.

Testimony

Dennis Andrew, New Stuyahok Traditional Council.

I'm a Fisherman from Bristol Bay. Last year fish went up river like a big ball. They used to be all spread out. Now fisherman start heading up river real quick. Next one went up Nugashik River was killer whales. Never seen killer

George Edwardson from Barrow, Alaska, President of the Inupiat Community of the Arctic Slope.

I've been fishing on the Ikpikpuk River all my life, ever since I learned to go out. 30–35 years ago, I used to put a stake in. Today I go to same spot, and the river bed dropped over 30 feet. The shape and depth of the river has not changed, but it has dropped over 30 ft. Another observation, kids growing up, would teach about the forest in the arctic world, raised them to understand those were trees. Took them to the same spot, and those willows now 7 ft tall, used to grow inch and a half. At Barrow radio tower, Japanese scientists found a black spruce tree 3.5 ft tall. Everyone knows trees don't grow in tundra; they are now at the coast of Barrow. There were seven robins in my yard. Birds that never grow that far north. And our rivers normally have white fish. Arctic whitefish are being replaced by salmon, moving in just last 3–5 years. Rivers are changing.

Permafrost now thaws to over 8 ft per year. Look at that sea level in 1970 sea level, 150 miles south of coast was 258 ft above the sea level. 10 years ago same thing, now 250 ft above sea level. Way up in mountains, talking about bedrock for ground, not bedrock or dirt, but sea level rising. Stories say that when all the ice melts in the north, my community will move to Umiat and we'll still be at edge of ocean. When looking at moving your communities, makes sure you're above 250 ft above sea level. This is not 1st time this has happened. This is the 7th ice age. These are normal geologic events but the contaminants make it faster. Sea level will rise another 250 ft by the time ice melts in the rest of the world. When you dig in ground you see sea life and shellfish. In Barrow so far in my life have moved back in 50 years over half mile from where beach house used to be. In 60's and 70's, reading taken at the point. With my dad hunting seals he was standing there looking back at the ocean. He said I stood in this same spot as a teenager, except I was about a mile out from where we're standing. He said I want you to check it. How do you check? These sailing boats took sextant readings from Barrow 50 miles down and 150 miles east. Got those GPS readings, figured out how to use it, was always wrong. Finally figured it out, shoreline had moved back a mile. The place we used to get drinking water is now in the middle of town. Animals are changing. Animals used to live in the Brooks Range are now in Barrow. That's what I've observed in my life so far. Thank you.

Wilson Justin.

I want to say about 250 ft level, I've been attending conferences for decades and have always wanted to hear a number. No agencies have ever come up with one. 250 ft. Thank you

Margie Hastings. Bristol Bay region, New Stuyahok Traditional Council.

My dad just turned 83 few days ago. He was an orphan but there were people willing to raise him. He remembers they would say that there will be changes. The universe will follow how the people on earth are. Elders are very perceptive. Elders follow the seasons very carefully. Used to wonder what does that mean? What kind of messages is he trying to give us? Wanted us to observe. We have to be observant. Know we're not like how we used to be a long time ago. A lot of us are pretty much disrespectful of our lands, our waters. We do what we want like ——— (Yup'ik word) peoples. It means people without regard to rules and standards. That's what the elders say at home. The way we are on earth, earth is becoming that way too. We've had some changes in Stuyahok area. My dad, he's a naturalist because he's an indigenous person. He's his own biologist, anthropologist, he's a scientist. He's got his own Yup'ik degree. He tells my nephews, don't go out tomorrow, stay home and chop wood because he can look at the weather and next day is a blizzard. These elders have own in their culture. He looks every year, I go out with him every year berry picking. He loves nature, that's where we get our love of nature.

Hills across our village ——- saying those hills across there are filling up with trees. Past 10 years I was there, I've been over seeing these hills out side our village. What used to be barren hills, —— - —. Makes me think of what he said —— ——- - —— our climate. I used to wonder how could that be? How could Alaska become tropical and low places where it's always hot become cold? But that's what he used to say. I've experienced it quite a bit. Extreme heat, when I was growing up it was always cold, always raining. But past few years it's like tropics. My Grandkids running around with only short pants it's so hot. — under sun, she gets nice brown tan sitting up there, sitting up there laying up there just like Hawaii, tan as she could be. There wasn't this kind of extreme heat. Touching anything metal would be hot. Kills our berries too, now after drought. Even in Alaska drought conditions with —, just like a hurricane in the tropics. We've experienced that several times in my community. Severe rain that's what elders are saying. Just like in the movies in jungle because they watch movies and they would say that we're going through change. These older folks, say Alaska and say Australia, people with warmer climates will have more colder climates and people in colder more tropical.

In our town call them ——-, but not sure of scientific name, Burbot? Kind of bottom fish. What they do is set out, son is doing right now, told him it wasn't right time, but he can do his own thing. — Have hook they baited and it sits on ground, comes and munches on thing and gets stuck, next day, ——-. In Stuyahok growing up, always had – burbot very white meat, liver is very delicious. Just know it as —, think it's burbot. Don't know what you guys call the —, in Stuyahok young men used to have hooks that just sit there. They're gone. Gone forever. Maybe because no burbot. Last year my son got maybe 2, and they were very small. ——- got only 2, gave to my parents. They hadn't had it in so long, they didn't even invite anybody, they ate it all themselves.

Also bumble bees. They don't look like the kind that grow on flowers. Grow on mud with sand. Kids are coming into classrooms with bumble bees like buds in their hands. Teachers say not, they'll bite and sting you. Whole bunch by elementary school. These bugs burrow in soft earth. There's a whole bunch. You can see because of the holes. I asked my dad and he's never seen this kind of thing. These bugs are unfamiliar to our community at New Stuyahok. That's one big change we never ever seen before.

Another is birds. Birds that kill good birds. They have really long tails. I asked my dad a few years ago. These birds started coming in, these birds are vicious, they come into our home and eat our beautiful swallows. They live even in 30 below temperatures, and these birds are flying, they can survive in extreme heat and cold. We've never ever seen these birds in so much abundance. Also notice these ravens, not as —, how do you say ——-? My parents said these ravens, if you even — once across room it would fly out. ——- They aren't like timid and scared of human beings like before. You see these birds won't fly away, just continue eating. Amazed my dad and mom, birds that used to be timid, not timid. These are some of things I've noticed, worked with elders last 10-15 years in college setting, ——- . ——

Eric Morrison, Douglas Indian Association.

Now I know where our ravens are going. We like you have a lot of climate change. My name is Erik Morrison from Sitka, now working in Juneau. Looking forward to seeing how these are written out in all your languages. Gunalsheesh. Worked in Sitka in charge of natural resources, in charge of inspecting rivers. Used to have a lot of sad days. We were so used to rainfall and heavy snow would come down and feed streams and salmon. But started drying out and salmon would be dying, see them in little ponds then nothing, they were dead. Glad they're showing up north. Our relatives can take care of them. We were worried about them. Everything is going through extreme changes.

Elders used to tell about berry picking. Sometimes no elderberries, blueberries, salmon berries, have to rely on Costco berries. Was sad thing for elders, not the same. Took some of youth out trying to teach them to gather traditional foods, trying to teach them how to gather sea asparagus. We could see they were mature, also infected, so said now not going to pick those. Is it climate change or change in times, pollution, China and India dumping all their pollution on Alaska. Other things really quite extreme.

For example, ever go to see, just look out and see certain area of trees where trees are decayed. Over 90,000 acres of yellow cedar died over last 10 years that we know of because of climate change. The snow used to be the blanket for the roots, it used to stay all winter. Now because of extreme changes it melts then freezes. Now about 90,000 acres dead, who knows how many more will die in the next few years. Maybe some of you up north can take our cedar because we don't want them to die.

Other things we have to adapt to like sunfish. What are sunfish, kind of look like the halibut but we haven't seen them before. Guess you can eat them, maybe we'll have to. Maybe they can replace the halibut that are disappearing.

Unlike you up north with land going down, in Juneau see change. Not surprised if before I'm done, won't be a change —-. Those are natural changes, without heavy glaciers land is rising. Many things going on. Sea turtles we've never seen before. Maybe we'll have to learn to eat those as well. Clams, another thing that's extreme to us. Elders used to be able to see Red Tide coming. Would have maybe 2 or 3 weeks when you could not pick, becomes poisonous. 2 people in Juneau died two years ago from eating them. Now can't see the red tides so readily, but they are all over. All of southeast now concerned about them. A lot of people that used to gather clams now don't eat them. Looks like all the way up to King Cove now. It's heavy and pervasive and taking away our seafood.

Abalone and sea cucumbers, result of people coming in saying hey, that's pretty good, now it's a commercial product for China and Japan, but also a result of climate change. Thank you for listening to me. Well Sitka where I'm from is probably last bed for herring resource. Still pretty plentiful but fisherman not seeing young ones, so very concerned about that. Young herring almost gone over areas like Ketchikan and Yakutat, almost gone. Hooligan the same. Maybe we can come up with resolve and force Juneau to create some boundaries to protect our resources and maybe they'll come back. —- well, clams are still there, but being poisoned by plankton harmful to humans, toxic and will kill us.

Tina Tinker, Native Village of Aleknagik.

July 13 of this year we walked out of home at 6 am to find ¼ inch of frost on 4 wheeler July 13. I was just so shocked. Even our skiffs had frost. Thick frost. So July 13 had frost. In area affected, some of our berries didn't grow this year. I looked at cranberry it was still a flower, should have been berry, so didn't get cranberries this year. Salmon berries grew late this year. I had my traditional foods, king heads and had to throw away 2 buckets because they didn't make the water, and we didn't want to eat them because of botulism.

And like Dennis Andrew said, had killer whales go up Nugashik. A few years back, go up to river to gather our whitefish. Slipped net underneath the ice. This time I didn't go up, but my husband went up. This one whitefish had this huge lesion on it, covered half it's body and up front. Decided to try to cut through to see what it was. Inside was just dark brown fatty tissue, didn't look normal. Should have shipped to Fish & Game but didn't know about that program. I am now decided to become an observer for Alaska Native Tribal Health Consortia, to try to observe some of changes we're going through, put pictures on website and where I found it.

When Margie Hastings was talking about rain, this year is the first year I've seen just a big down pour. This summer around July 3 or 4, drops so huge they'd hit ground and bounce up about a foot. First time I ever seen that. Every year we're the first people to get moose, this year we didn't get one. Just had season to catch 2 moose for winter, usually catch 2 to feed family and share with elders. This was first year elders had to share with me. Bulls are staying way inland while females are out grazing. I think that's another part of climate change. Thank you.

Frank Pokiak, Invialuit Game Council.

Thank you. I think I'd like to say a few words about climate change. I come from Canada, little community called Inuvik, right by the Beaufort Sea. My name is Frank Pokiak. My parents came from Kaktovik, I probably have relatives out here I don't even know. Live right by Beaufort Sea. Where I live, we're always having problems with erosion.

20 years ago we heard about climate change. Looked at highest hill. 20 years ago we were hearing community better move houses to higher ground.

On species, we also observed killer whales and different types of birds we never saw before like turkey vulture, blue herring. Species like that are coming up north. Seasons are really different now. I grew up traveling and could pretty much predict what weather was going to be, and really can't do that anymore. Listening to weather forecast, gets very dangerous for our people harvesting. Listening to forecast and observing ice conditions, didn't have to worry. But ice conditions are thinner now and create problems for our harvesters. Our seasons are really changing.

Used to hunt geese, like to hunt when heading for their nesting ground because really nice and fat. Used to stay out there until 10th or 15th of June without Skidoo, but can't do anymore. It's melting quicker. Really late freeze up and early breakups. I think we have same observations and concerns as a lot of you, but thought I would share coming from a different country. Thank you.

Juanita Wilson from Northway Traditional Council, located up near Tok.

I'm from the interior, not by a coastline. I noticed changes, some of the same changes on bees she was talking about last summer. A lot of bees and bee hives around my house. Before there were hardly any bees. Another thing I wanted to talk about, our rivers are also eroding. The road we take out, now water flows over the road. We had to move it back, and will have to again now. Another thing about berries also. I remember last summer and the summer before, can't find blueberries anywhere. I remember every year as you see first blueberries, now we can't find them. Now we have to go all over to fill your bucket. We're also noticing similar changes as you see on the coast. After today I think I'll be more observant. There are a lot of changes, and we're also experiencing them. Thank you.

Julia Dorris from Native Village of Kalskag, from the middle Kuskokwim.

Thinking about people and behaviors, especially of animals. We have black and brown bears, and in November flying between Kalskag and Aniak, just saw a brown bear and grizzly bear. Had a lot of snow, would think bears would be hibernating. Never heard of Brown bear running around in winter time. It was cold too. Even in summer have different butterflies then we ever saw before. Different ones and different colors, big ones in fish camp.

Berries, my cousin went out, usually there's lots but he just had to walk all over and didn't even fill the bottom of a little bucket. All summer long there seemed to be a lot of black bears coming into the village behind our house. Summer black bears not scared anymore, come into the village. This never happened before at houses, only fish camps. A lot of wolf sightings, running down road in middle of Kalskag, three of them. Usually shy and stay away from humans. Moose are getting braver too. Lot of moose in the tundra near the coastal area of Kuskokwim. I called my daughter and said there's moose behind the house. In the housing area, there's a lot of houses, thought only city moose in Anchorage were that brave. Now village moose are getting brave.

Mike Brubaker, Alaska Native Tribal Health Consortia.

I want to invite everyone to attend a session tomorrow on the Local Environmental Observer program. Tina mentioned local observer program, just started enrolling people two weeks ago. Provides a tool for tribal staff working on environmental issues to post observations onto www on map and capture and share and learn about changes happening and ways to adapt. Tomorrow at 9 am doing introduction to local environmental— progress, how it works, and provide us with feedback on how to make it work better.

Carol Oliver from Golovin.

We have a lot of the same issues in our village. A lot of bears, they're more in groups and there are also more. Never used to be afraid to go out berry picking. Now if we do go out, we need to have a guy with a gun. Seeing a lot of sink holes because of climate change. A lot of erosion. As soon as it melts, we'll take some pictures to share. Doing a lot of damage. This winter is first time somebody caught a coyote in our village. When they caught it everyone was taking pictures. I said what? A coyote? Don't know how far north people catch coyotes, but that was a first for us, that's unusual. Thank you.

Wilson Justin.

Just as a final thank you before I step off, really want to thank you for your participation.

Elizabeth Asicksik from Tununak, on the Nelson Island coast.

The river is changing very rapidly and causing erosion, fish are changing, seals are changing and getting sick. Thanks, that's all I wanted to say. Thank you.

whales go that far up river before. Snow witnessing today, now, talking a lot of snow. Hear about Valdez record snowfall. These are the changes we're seeing. Thank you.

There are two sign up lists making their way around. Please be sure to sign up for technical information and resources for climate change we'll email you.

Note Re: Methods.

This session was recorded by two typists in the audience, as well as in hand-written notes. Original files and copies are available on request. There are audio tapes, but those are not available yet. The notes were carefully transcribed and reconciled, to be as accurate as possible to the testimony given. We regret any transcription mistakes and are grateful to those who traveled so far to share their observations and knowledge.

Gunalchéesh,
Michelle Davis, Tribal Coordinator
US EPA, 222 W. 7th Ave., #19
Anchorage, AK 99513
(907) 271-3434
davis.michellev@epa.gov

This page intentionally left blank.

Appendix D. Climate Change in Tlingit Aaní

By Kanaan Bausler

> The earth was covered with flood waters and the Raven stayed above the water… all the land was covered with water. His beak was stuck in the clouds and he stayed there. It held him fast; he stayed there until the water started receding. It let him go and he started coming down. He had no idea where he was. He made a wish when he was coming down, and he said, 'I hope I land in a safe place. I wish I land in a safe place.' And he came down on a little island up near Yakutat. That island's name was Gajaa. Landed on a small sandy beach. And when I talked to the Eagles, I told them that 'Ravens wish when he was coming down, is our wish for you and your sorrows. That you come to a safe place, where you can find comfort. Where you know that we will be here to help you whenever we can.

—Seitaan Ed Kunz, L'uknax.ádi

The wish to find a safe place to land is likely a familiar feeling to many Tlingit people today. A culture that has survived centuries of disorienting challenges, the Tlingit Nation is steadily regaining balance. The native peoples of Southeast Alaska are constantly finding new ways to integrate cultural traditions into modern American life (Hope and others, 2000). The discontinuation of pre-contact lifestyles inevitably loses many aspects of the old way of being, yet the transformative power of colonialism is a reality. This reality defines the conditions that the Tlingit people must adapt to in order to retain their culture. A significant part of these adaptations involve communication between different cultures. Although the Tlingit people need to be able to evolve with the global humanity, community members from all walks of life also need to be able to appreciate Tlingit perspectives. By exploring Tlingit perceptions of climatic change one may discover unique insights into the connections between human beings and environments, and the implications that environmental changes have on past, present, and future generations.

One form of communication that is steadily improving is the realm of academic knowledge. Researchers and scholars are consistently giving more attention toward the accumulated wisdom of oral traditions. Authors such as Shari Gearheard (Gearheard and others, 2009), Nancy Turner (Turner and Clifton, 2009), Wade Davis (Davis, 2009), David Abram (Abram, 1996), and Julie Cruikshank (Cruikshank, 2001, 2005) have spent decades around the world searching for the interface between western and indigenous knowledge systems. The improving interactions between academics and natives are allowing both groups to benefit from one another's observations.

> Fish and game and the forest service got a hold of the native corporations and said, you know we know the forest, we can identify every plant that's out there, but we still don't know a lot of things. The elders know that. They grew up with it so there's a lot of information there that we need to know. I'll give you a quick example. Three years ago, when they knew that the yellow cedars was being killed off… They realized that the roots were shallow, so that they'd freeze and that was what was killing them. They dug up all these yellow cedar saplings and started an environmental experimental forest out at 40 mile. The way they did it is they put some low, some high, and some on the side, but did it in an open area, so that it could get good heavy snowfall that could protect the roots, and then go back every year and take students and document all the stuff. And what they did was, every sapling that they brought from no matter where it was, they were all logged, tagged, dated, everything. Showed where that little sapling came from… It was only this last summer that I brought the subject up to another elder, who is older than I am, and he knew the forest from hunting and what not. I told him about the experimental forest yellow cedar project going on out the road, and his question to me was, 'so when they dug up that little sapling, did they plant it the same way that they dug it up?' And I asked him, 'tell me what you meant when you asked me that.' He said 'well a tree, when it first sprouts up, its branches, some will be north, some will be west, east, south,' and he said they have to plant it the same way. I took that information back to the Forest Service and they said 'we've never heard of that!' And I said 'well I didn't either it was just brought up by an elder that I happened to be discussing this with.' So you know there's all kinds of little things that you have to, you know…

—JO-OUAACK John Morris, Yanyeidí

This incorporation of oral wisdom into modern academics has been given the label "traditional knowledge". But what qualifies as traditional knowledge? It's almost oxymoronic to try to define this term, as its meaning goes beyond the ability of written language to describe it. As author Julie Cruikshank has noted,

> Narrative recollections and memories about history, tradition, and life experience represent distinct and powerful bodies of local knowledge that have to be appreciated in their totality, rather than fragmented into data, if we are to learn anything from them… Modernist recasting of 'traditional ecological knowledge' continues to present local knowledge as an object for science rather than as intelligence that could inform science.

—Cruikshank, 2005

Nonetheless, in order to write about traditional knowledge, a description is needed. Perhaps it is the indigenous way of interacting with the world, the cultural way of life. Traditional knowledge demonstrates accumulated generations of experiences with a particular location by a core community of people.

> I grew up on the Taku River until I was 15 years old. I know the Taku like the back of my hand. I used to tell people, if you took me and blindfolded me and stuck me anywhere on the Taku River, and left me for an hour or so, I could tell you where I was at, just by the fragrances or the sounds, and what season. I'm that familiar with it. I LOVE THAT AREA.

—JO-OUAACK John Morris, Yanyeidí

These layers of experience have been passed on to younger generations through oral sharing methods, although modern groups have also incorporated written techniques. Traditional knowledge differs from western written methods in that the knowledge is not stored in a permanent medium, but rather in the flexible memory of the living culture bearer. This passing of knowledge can come in many forms, such as direct descriptions of observations:

> It was very interesting what occurred one day when this man, Dáanaawaak, called me up and told me he had something he needed to talk to me about… And he spoke to me in the Tlingit language.
>
> He was very very very concerned. About our earth. Haa Aaní. Tlingit Aaní.
>
> He wanted to tell me the way he observed things happening to the earth.
>
> My grandfather Dáanaawaak said we are the ones who are to watch, to care for, to observe the earth and how it is being cared for.

—Kingeisti David Katzeek, Shangukeidí

Or from clan stories:

> We would all, at the same time in the spring time, last day of school, load up the boats and go to the Taku. Build our cabins, tents if we had the fish camps, berry camps and what not. And since I was the oldest, I was his (John's grandfather's) favorite. So he would call me over to sit down on his big rock and he would tell me stories, that's how those stories got passed down.

—JO-OUAACK John Morris, Yanyeidí

There are clearly many ways to pass on traditional knowledge within a family. But how can these ideas regarding the nature of existence be shared across cultural boundaries? This is an extremely difficult question in that it requires one way of knowing to be described in the terms of the other. How can one write about an oral language with out missing something integral? Inconsistencies are inevitable parts of the process. However, for those seeking to understand the world, those that are searching for truth, it is a necessary process. To be able to see a problem from multiple perspectives is crucial for successful solutions.

> Traditional and scientific observations are independent sources of information that can be brought together to increase confidence and depth of knowledge. While both methods have uncertainties of their own, overall uncertainty can be reduced when the methods are combined. The purpose of such comparison is not to "validate" one set of observations in terms of the other. Rather, it is to combine them while taking advantage of their differences.

—Huntington and others, 2004.

If each form of observation may benefit the other, then what are some general contributions that these sources can make to each other? Academia can provide indigenous intellectuals with structured methods of determining precise cause and effect. With constant change in the world's ecosystems, specifying the causes of a conflict is crucial to the management of results. Traditional knowledge, on the other hand, may provide western scholars with a humanistic approach toward these problems. In the western academic eternal quest for objectivity, a hole develops, begging the question: where do my community and I fit in to this? Traditional knowledge methods embrace subjectivity and often seem to blur any lines that might separate the self from the others.

> We're so connected as human beings to our resources. We were just talking about one Lifeway that my brother Michael recorded on a silkscreen of a woman's life and how we marked her life through this story of a child becoming a woman under the tree. And it's such a powerful symbol in that the moon times when a woman becomes a woman, she has the moon times every month. And he has the moons around this tree, a big spruce tree where the women are taken to have their menesis, or before they have their menesis. To learn how to be a clan member. To learn how to be a woman either sewing, weaving, or being a mom, or being an auntie. Learning those ways under this big tree. It's such a powerful symbol because not only is that tree still standing… with times changing, our resources are seen differently too. Which eyes are you looking with, the older eyes or the newer eyes? Things that were there we don't see anymore but it's still there.

—Kasts Saa Waa / Stalth Kaa Waas Della Cheney, Kaach.ádi / Tiits.git.aa.nee

In my research attempting to document traditional knowledge into a format that is accessible by all parties, I became increasingly perplexed by this challenge. How can I share this information in a way that respects both knowledge systems? How can I reduce the distillation of a potent source of wisdom, knowing that the final product will include a written report? In order to effectively express it, I had to impose categories. However, taking a little bit from both styles, I have chosen to present the information from my subjective experience. The remainder of this report will follow my experience with collecting this information, and my interpretation of how to best proceed with this project.

The original purpose of this report was to submit Tlingit "indigenous and tribal technical inputs" to the National Climate Change Impacts Assessment, a United States government document released every four years by the U.S. Global Change Research Program. I started by compiling a list of people to talk to and literature to read. I did a literature review to compile resources, I attended relevant conferences, presentations, and gatherings, and I interviewed Tlingit elders. As expected, the richest source of information came from the interviews. After recording conversations with eight different elders, it was apparent that these recordings would be the driving content behind the report. I transcribed each interview, filtering for climate change related passages for the sake of time. After transcribing all eight, I determined the themes that ran throughout the interviews and established eleven categories that I could assign to each passage. I then made a directory of the interviews, showing which categories were described in each interview and where to find them on the time code.

The categories that I chose to label these interviews with are as follows: Introductions, References, Physical Change, At.óow, Incoming Species Migration, Species Loss, Lost Traditions, Challenges to Adaptation, Adaptation Techniques, HiStories, and Wisdom. Introductions label the places in the interview where the speaker talks about him or her self and where he or she comes from, while References labels places where the speaker suggests other sources of information. These first two categories are not discussed further in this report but can be found in the interview directory. Physical Change labels observations of non-biological matter transformations. At.óow labels references to clan properties that are connected to stories of environmental changes or challenges. Incoming Species Migration labels observations of environmental change in the form of new habitats and residents. Species Loss labels observations of environmental change in the form of the disappearance of habitats and residents. Lost Traditions labels cultural activities that were referred to in the past tense and no longer happen on the regular basis that they once did. Challenges to Adaptation labels systems or events that have deterred local people from being able to efficiently evolve to imposing changes. Adaptation Techniques labels programs or actions that are improving the ability of local people to evolve to changes. HiStories labels narratives of Tlingit history with a special emphasis on stories (I chose to make a new English word for this because the words History and Story do not convey the meaning that I implied. History is not specific enough and Story seems like a condescending suggestion of fiction). And finally the last category, Wisdom, labels philosophical observations, theories, and advice from these deeply intelligent culture bearers. The remainder of this report will explore each category, demonstrating samples of the information I found applying to each, and providing suggestions for further research.

Physical Change

For the Tlingit of Southeast Alaska, climatic cycles have defined the elements of life since time immemorial. This region of the world has been particularly influenced by changes in the global climate. The characteristics of the region make it very vulnerable to the fluctuations of the climate. The northerly, but sub-polar latitude makes Southeast Alaska consistently proximal through geologic time to the coastal region where water freezes at sea level. As the cooling periods allow for large-scale glacial advances, and warming periods bring the regional retreats, the livable land for humans is tenuous at best.

Modern Tlingit continue to observe changes to the physical environment. The region experiences such high levels of physical energy with the intensity of the weather, and being so close to the hydrological melting and freezing range. This allows changes on the geological scale to be observable over lifetimes, as opposed to other regions, where climatic change is not as obvious. In my interviews, many individuals pointed out physical changes that they had noticed. Among the most common were a decrease in the amount of glacial ice in the ice fields and waterways, an increase in air and water temperatures, a decrease in snowfall accumulation, terrain uplift from isostatic rebound, and variable effects on fluvial discharge. Here is one example of an account of changes in snow accumulation:

> And then the snow. The kinds of snow that we get are unusual. We got tons of snow a long time ago. I mean tons of snow. You almost wondered when it's going to stop snowing. But now, you know we get periods of snow but we don't get the kind of snow that we got in the past.

—Kingeisti David Katzeek, Shangukeidí

Decreases in snowfall are typical for many coastal regions at sea level, as the warmer temperatures bring rain with storms that used to come as snow. The resulting retreat of the accumulation zone in glaciers can cause a decrease in ice quantities around the environments where humans spend most of their time. The following is an observation of the effects of this process.

> Even during the 1940s and 1950s we would see icebergs coming out of the Taku River and they would float right up the [Gastineau] Channel. And all over. Going down south and around the backside of Douglas Island. Huge, huge icebergs. We don't see that any more. And the reason we don't see that any more is because the Taku Glacier is actually right up against a great big mountain right here and it sure is sloughing off into the river but what happened is, as the Norris Glacier retreated, left great big sand bars where willow and seed grasses grow now, which put up a buffer system so that this glacier… icebergs aren't going into the river and coming down.

—JO-OUAACK John Morris, Yanyeidí

University of Alaska Southeast professors Cathy Connor and Daniel Monteith worked with a team to publish a multidisciplinary paper entitled "*The Neoglacial landscape and human history of Glacier Bay*". In this study, they took Huna Tlingit stories of the Glacier Bay region and combined them with radiocarbon dating from geologic evidence to recreate the historical landscape and model the extent of the ice from past centuries. In addition to being a great example of sharing knowledge to create a valuable product, the paper paints a picture of how the bay has changed over time, as well as how human habitation has been a product of the physical environmental conditions. It demonstrates that despite the challenges of living so close to such powerful forces of nature, the Tlingit inhabitants thrived and developed an extensive history with the place.

> Even the Chookaneidi in Glacier Bay, they lived there with the glaciers very close. Even in our historical songs we talk about people passing under glaciers so I think they certainly had a respect, but not a fear of glaciers. They accepted them as part of their environment. They weren't something that we look at from afar like we do today. They were an active part of being.

—Koonesh Eric Morrison, L'uknax.ádi

A major source that the research team used in the study was the Glacier Bay History, one of the most well known Huna Tlingit HiStories that has been published and widely shared (Dauenhauer and Dauenhauer, 1993). In the HiStory, a young woman commits a taboo by teasing a glacier, Sit'k'i T'ooch' (Little Black Glacier). This influenced the glacier to surge rapidly, taking out the village and forcing the community to relocate. This is certainly an incredible HiStory, and it brings to light a common theme in other well-known HiStories. Being forced into migration by changes in the physical environment, and particularly by ice, are well-recorded events due to the nature of how Tlingit clans are named.

At.óow

At.óow is a Tlingit term that refers literally to "owned things" or clan property. The at.óow concept can apply to material objects, places, names, stories, and other cultural elements. Gooch shaayi Harold Jacobs, Yanyeidí, is an excellent source of information on at.óow and the HiStories behind them. Please refer to my interview with him and his Facebook photo album of Tlingit Place Names for more information.

Clan names are a form of at.óow, and a particularly valuable form for tracking climate change. Tlingit clans are often named for the places that they formerly inhabited. With this concept in mind, one can investigate clan names in order to determine migration histories and perhaps the events that influenced the migration. In my past research of Tlingit place names, I helped X'unei Lance Twitchell, L'uknax.ádi, with his work in cataloging these clan names and tracking origins. An example of this is the Naasteidí, the clan "Belonging to the Nass River".

Interestingly, almost every major river that divides the mountain ranges connecting the interior of the continent with the coast has a migration story involving glaciers attached to it (Cruikshank, 2001). The Nass, Stikine, Taku, Chilkat, and Alsek are the major fluvial highways of migration and each has its own clans that are connected to them. It seems that in coming to the coast, some of the original Tlingit clans had to navigate to the other side of a glacier that was blocking the path. A commonly told HiStory comes from the Stikine.

> My dad's clan has a story of a migration under a glacier. They were migrating, and got stopped by a glacier. I don't know how long they were there. But when they decided to move on they sent a man over the top to see where the river came out… He told where it was so they sent two old women under the ice. One was named Koowasix, which means 'besieged' or 'stranded'. And it seems to come from when they stopped at the back of the glacier. She may have been a baby when they stopped there. The other one's name was Aawasti. They put them on a raft, they say canoe but I think it was a raft. They put tree boughs on their head to see how wide the tunnel was under the ice. They came out safely on the other side. They decided to, they went back and told the people it was a safe passage. Some of the people came out under the ice, others didn't want to go under and went over the ice. The clan that came out under the ice wound up in Tongass and Angoon and the group that went over the ice wound up in Chilkat. They're the same clan, they have two songs that go with that story.

—Gooch shaayi Harold Jacobs, Yanyeidí

Perhaps the time period in which Tlingit people were migrating from inland was particularly cold, influencing the continental groups to seek the warmer climates of the coast. Even if each group came individually and in different centuries, the individual causes of migration could potentially have been climate related. The fact that most if not all of the major rivers have a glacier encounter migration story attached to them seems to support the idea that the times of migration were glacially active in the advancing direction. Here is another one from the Taku:

> The story is that when the glacier had come clean across the river, the Taku valley, it had sharp cliffs that come down and the glacier came over and covered the whole river. And at the time, the T'aaku Kwaan people, all they knew was the glacier and the ice. They didn't know what was on the other side. They didn't know what was over there. So the chief every summer when they went up fishing and berry gathering and what not, would, he would designate some warriors to go over the glacier and see what was over there. To look. And every time they did they would come back and say that 'There's nothing over there. Everything's nothing but snow and ice, no matter how far we went it's nothing but ice and snow.' So it was taken as fact. That once you got over the glacier there's nothing there forever except ice and snow. But one summer, while they were up at camp, a carved paddle floated out from underneath the glacier. And the people saw the paddle and said, 'There is somebody over there. There are other people there.

—JO-OUAACK John Morris, Yanyeidí

Unfortunately this story got cut off by a phone call in our interview, but luckily a version of it has been published in *Gágiwdul. àt: Brought Forth to Reconfirm: the Legacy of a Taku River Tlingit Clan* by Elizabeth Nyman and Jeff Leer (1993). In "The history of the T'aaku Yanyeidí" the coastal tribe crosses the glacier to meet the group on the other side. After making peace, they drip blood on the glacier, influencing it to open the river passage, thereby connecting the tribes and allowing them to be one people. Yanyeidí means "Mainland People". Another clan name from the Yanyeidí group is S'itkweidí – "Belonging to Among the Glaciers".

Another very descriptive form of at.óow is place names. Just north of the Taku, the Mendenhall is another glacier that comes down to sea level. Many of the interviewed elders referred to the retreat of this glacier as an observed physical change. A nearby place name is particularly telling of the past environment. The Tlingit name of what is called Montana Creek in English is Kaxdigoowu Héen – "Going Back Clear Water". This name can be traced back to a relatively recent time period, when the Áak'w Tlingit were occupying the area. Based on conversations with Richard Carstensen, who has done personal research on the history of the area, the Áak'w Tlingit had only been living permanently in large numbers in the Auke Bay/Mendenhall Valley area for a few generations before being joined by European-Americans. This presents a relatively small window for the time period in which the name was assigned. The name Kaxdigoowu Héen – "Going Back Clear Water" suggests that the mouth of the creek, where it meets the larger glacially fed Mendenhall River, was silty. Today a very clear line separates the opaque glacial river from the transparent forest drainage. The Tlingit name implies that the transition from clear water to silty water was less defined and farther back from the confluence. This indicates that the area was recently a glacial environment of freshly chewed rock that had not yet eroded and washed away, muddying the waters in the glacial valley.

More examples of place names as indicators of past environments can be found in the Connor and others (2009) paper about Glacier Bay. This study made extensive use of place names in their reconstruction of the past landscape.

> The interpretation of the Tlingit homeland hinges on 11 Tlingit toponyms that anchor two enduring oral narratives… S'é Shuyee (Area at the end of the Glacial Silt) indicates…that the people preferred living farthest from the direct effects of that hostile environment, and closest to, or with reasonable access to, tidewater… Interestingly, the stream that flows into the modern Bay just north of Rush Point is called Chookanhéeni Yádi (Child of [Grassy River]), a name that implies a secondary rank, such as a tributary to a larger stream.

—Connor and others, 2009.

Much research has been done on Tlingit place names and their capacities to describe past and present interactions with locations. A monumental book was released as this report was being written. *Haa Léelk'w Hás Aaní Saax'ú: Our Grandparents' Names on the Land*, edited by Thomas Thornton (2012), documents hundreds of place names for each region of Southeast Alaska. This book is essentially a library worth of information in itself, and will surely provide a deep source of information for interested parties for years to come. For more information I highly recommend other work by Thomas Thornton, particularly his book *Being and Place Among the Tlingit* (2008). As a final example of place names and environmental change, we will turn back to Gooch shaayi Harold Jacobs, Yanyeidí.

> There's a bay down in the south of Sitka called Necker Bay. There's a species of sockeye there. Sockeyes are called gaat. But the Necker Bay sockeyes are smaller and they call them dagák'. And Thimbleberry Creek Road, you know where Thimbleberry Lake is? About 2.5 miles out the road I think. The creek there is called dagák' yateen héen, which means you can see those little sockeyes in the creek. Apparently those have died off, I have never seen a sockeye in that creek. But they used to be there. I don't know what happened to that.

—**Gooch shaayi Harold Jacobs, Yanyeidí.**

Ecological Change

The primary observations of environmental change that I recorded were changes in species behaviors and habitation. In categorizing the interviews, I broke these observations into two different topics, *Incoming Species Migrations* and *Species Loss*. I will present them together here because they are so interconnected. However, these are such large topics that only a few examples will be shared.

As the global climate warms, northern habitats have become accessible to species that were previously contained to more southerly latitudes. Polar species migration has been particularly expedient in the oceans, as small changes in water temperatures can open a whole world of possibilities to aquatic creatures. Additionally, the fluid nature of the oceans allows for direct movement to wherever the conditions are best fit for the being. The process is a bit more challenging for land dwellers due to the excessive fragmentation of the landscape with barriers imposed by humans. As the Tlingit are a sea faring people, observations of marine ecological changes are thorough.

> They asked me to come and take them to North Douglas. And that's the first time I saw millions of those little muscles. Yein. Yeah they call it a sea cucumber. But yeah millions of them on that point, North Douglas, I was amazed! I didn't think they grew up here! I knew that they were being harvested around Hydaburg and Metlakatla. North Douglas... I never went looking for yein before, but there they were! That really surprised me. So think about this, if they were usually down around Hydaburg... that's where they were, what are they doing up here? I never expected to see them. I kept thinking about that and I'm telling you about it now. Never seen them up here. And it's a lot warmer around Hydaburg, where all the little islands are, and Metlakatla.

—**Kaayistaan Marie Olson, Wooshkeetaan**

> Species migration from shellfish to seaweeds that weren't identified any further north than Oregon were now showing up on our coasts. Tropical fish like Sunfish were starting to show up in our waters in the summer months.

—**Koonesh Eric Morrison, L'uknax.ádi, interview**

> What are sunfish? They kind of look like the halibut but we haven't seen them before. Guess you can eat them; maybe we'll have to. Maybe they can replace the halibut that are disappearing."

—**Koonesh Eric Morrison, L'uknax.ádi, testimony at the Alaska Forum on the Environment**

Although changing at a slower pace, terrestrial habitats are also being infiltrated by new species.

> I fished Glacier Bay through the 60s. There was never any moose there then, the big thing that the people from Huna went in there for was mountain goat. But never any discussion of moose out of that area.

—Koonesh Eric Morrison, L'uknax.ádi, interview

While some species are moving in to Southeast Alaska, the accounts of ecological change are overwhelmingly in the minus. It is very clear that the region is far less plentiful in life forms than when it was Tlingit Aaní. Many species populations have been diminished substantially over the past century or two. Every individual I interviewed had lists of species that have taken significant hits. Here I will share two marine examples and two terrestrial examples that demonstrate the impact of losing culturally valuable species.

The dynamic physical environment of Southeast Alaska has always been a determinant for the behaviors of the people. Not only has it controlled where they have lived; it defines their subsistence activities and ability to harvest. Perhaps the most significant creature to the Tlingit way of life is the salmon. Salmon are the keystone species of the ecosystem, and the lifeblood of the culture. Any changes to the salmon are sure to be noticed, and the changes over centuries have been numerous.

> How is it affecting us? It's affecting us very, very greatly. In my lifetime, we knew exactly almost to the day when the king salmon would come back, you know like the Fifth of May? Everybody knew it. We all knew it because we did it every year all our lives. All of that was handed down through generations and generations, not put down on paper but through word of mouth, oral. Oral history. And so we knew when the king salmon would come back, we knew when they would go in to spawn. How did we know that? Because of the temperature of the water. The temperature of the water had to be just right. They would go out, and say if they were going to go out Berners Bay, up Berners River, or the Alsek, or out at Yakutat or the Taku River or the Stikine, they would all congregate at the mouth of the river. Until the temperature of the water hit a certain degree and then they would go in to spawn. That was their signal to do that.

—JO-OUAACK John Morris, Yanyeidí

> We've seen pretty big floods because of the heating, the heat that is bearing down on the earth up in the Chilkat area, and the ice melting and filling up the river quite a bit and it's had its share of damage with respect to the salmon people. And if it hurts the salmon people then it's going to hurt us. Because we're just like – salmon, need oxygen, they have to have oxygen, they live off of oxygen. And spring water is one of the places where they go to get, its called Ishkahéeni, there we have a word for it right now. I mean this is thousands upon thousands upon thousands of years of knowledge. That the salmon go there
>
> The oxygen for the salmon is there
>
> We learn that the water that is the coldest, and the coldest is spring water, has the greatest amount of oxygen, any scientist can check off what I'm saying and they're going to find out what I'm talking about and saying, and what the elders have been saying for thousands upon thousands of years is true. So there's got to be some recognition. There's got to begin to be some acknowledgement with respect to indigenous peoples' understanding of the environment that we live in."

—Kingeisti David Katzeek, Shangukeidí

A direct consequence of the lack of understanding can be seen through another fish, the herring. Though practically everyone I talked to mentioned something about the herring, two stood out in particular:

> One thing I have noticed in Sitka is, commercial fishing with herring, they'd usually spawn from Hayward Straight all the way down to Dorothy Narrows. The channels and inlets and beaches all the way down would be white with herring spawning. It's just scattered areas but some how the State Department of Fish and Game figures they're producing a healthy fishery. Which I wrote about, I can give you an article that I wrote about that being the biggest want and waste fishery ever. And what are they going to do when they wipe out the herring? And the salmon and the mammals that feed on the salmon and the herring? It's having a major impact on the Sitka area. And so now when it looks like there might be problems, who do they blame? The subsistence users.

—**Gooch shaayi Harold Jacobs, Yanyeidí**

> As long as I can remember our people used to go out there to get herring, you know set branches out there for herring eggs. And the Department of Fish and Game took over matters in that area, and within 10 years it just disappeared. It's not there anymore. One of our friends, one of our elders, he was walking on the beach around out there with a friend of his and it was low tide and they were talking about changes that were going on. And they were busy slapping on mosquitoes, and one of them said to the other, he said, get the Department of Fish and Game to manage the mosquitoes around here. And in ten years they'll be gone.

—**Seitaan Ed Kunz, L'uknax.ádi**

While this conflict may have more to do with mismanagement than climate change, my time with these elders has taught me more than anything how interrelated these problems are. Whether human -induced environmental change comes as a consequence of over-harvesting or pollution, it all comes back to the same issues of communication and respect for common resources. These ideas will be revisited later in this report, but the herring observations bring up a good opportunity for further research. Is the decline in herring stocks purely an exploitation problem, or are other environmental and climatic changes playing a role in amplifying the depletion? There are many questions yet to be asked. Including questions about other culturally significant changes, such as the decrease in berries. What would cause an entire region to come up short on berry harvests?

> Last year there were no berries. No, no berries. We had to go out to Fred Meyer and they had those Oregon berries, or the berries from South America. Not true blue berries, but you call them blue berries… I don't know if you've ever been to a party, a memorial party, but the berries hold a special part in our culture. The food is served, and an announcement is made that this is the favorite dish of so and so, whoever's memorial we are at. When it's berry time, they hire people from the honorees guests, to come out and get paid for this thing to come out, and they bring the berries out with a song and a dance, sing and dance bringing the berries out, so it's a pretty big thing at the party. And last year there was no berries… I've heard stories. They were like that once before. And somebody said it was the bees because we never saw any bees around.

—**Seitaan Ed Kunz, L'uknax.ádi**

Having access to experienced field practitioners is about as valuable as any other resource for conducting research. This is why traditional knowledge is such an important thing to appreciate. These people have lifetimes of experience to draw from, to inform the rest of the community of what is abnormal, what we should be paying attention to. Unfortunately this information can sometimes come at the expense of important elements of their culture. And what's worse, at the expense of future generations.

> Another thing that's popular for us to teach the kids like gathering berries, has become a haphazard hobby because you never know you almost always had berries at some point during the summer but now it's a guess whether you're even going to get any of these berries like that. We used to have historically like huckleberries, it's hard to find now in the summer things like nagoon berries even salmonberries and blueberries it's very sporadic. You can still go to certain areas and find blueberries but its not healthy like it used to be. You always used to have a good crop somewhere and could depend on it but now you don't know because the weather even in the summer months is hard to predict what's going to happen.

—Koonesh Eric Morrison, L'uknax.ádi

Perhaps the loss of berries is a climate related issue, or perhaps something else is the primary cause. Either way we can see that it is an ecological change that comes at a tough price. Another species, the yellow cedar tree, is of equal significance to the Tlingit in their traditional production methods. As the decrease in snowfall reduces the amount of insulation for the shallow roots of the yellow cedar, the big freeze periods in winter are able to kill the trees more easily (Hennon, 2008). The loss of this culturally valuable tree is perhaps the most consequential directly-climate-related ecological change.

> The other thing that we are noticing now is that the cedar trees are struggling, the yellow cedar especially... With hearing about the yellow cedar and how much I use it for weaving and how hard it is to harvest the material now, as I get older, I realize how important it is to find out what's happening with the tree… So it is a concern for us to hear about what's happening to the cedar trees. Right now it's just the yellow cedar that they're talking about but whatever affects something affects everything else so… Some of the other things that I find, and man is part of the problem, man generic, meaning man and woman. In our environment, as times change, all these things that bring us immediate satisfaction is part of the problem because the time it takes to learn how to process something takes time away from these things for our younger people. It's harder to attract their attention.

—Kasts Saa Waa / Stalth Kaa Waas Della Cheney, Kaach.ádi / Tiits.git.aa.nee

In looking at ecological change we see a culture that depends so heavily on the food that they eat and the materials that they create with. It's hard to imagine the Tlingit without salmon or cedar. Yet it's a reality that they have been struggling with for generations now. The relative losses over the years are as significant as any could be. The progressive transition from abundance to depletion is not as obvious to the average modern resident, whose perception is likely on a shorter time scale. It's hard to believe that the next step in the process is regional extinction, but from the perspective of a Tlingit elder, the transition has been as swift as a lifetime, which, on a traditional knowledge timescale, is not that long. Logic would tell us that things are getting worse. And for the Tlingit, the losses carry over to all aspects of life.

Lost Traditions

Food is everything for any creature. Any human. Any culture. The way that we survive on this earth is a product of how we secure energy for ourselves. For European-Americans, the system of agriculture has allowed the removal of the individual from the energy capture process. Western societies can depend on others to produce food, thereby allowing citizens to define culture with other terms, other relationships. However, it is important to appreciate that it is an agricultural culture, and the concept of the farm is one of the most fundamental elements that defines the way of life. For the traditional Tlingit way of life, food harvesting was the responsibility of every member of the civilization. Therefore everything revolved around the energy capture process. The loss of culturally significant species then, is more than just plants and animals, more than just food, more than just tools. If the Tlingit people aren't harvesting, they are losing a fundamental part of being Tlingit.

You see, my wife and I, we put up a lot of our food. And we had to go up, we went out to Fred Meyer to get our sockeye. To get Copper River sockeye shipped down, so we go out there to get it… Percy and I used to go up to Haines every summer and we'd stay up there for a month, two months. And we'd put up all our fish. We'd fish in the Chilkat River and out at 8 mile, 8.5 mile we'd have a camp out there. We'd just camp out there for 3-4 days, come back into town and smoke some of the fish. Put others right in jars. Whatever we did Percy and I, we'd split right down the middle with her sister. She was the one who was teaching us how to do these things. If we weren't fishing we'd be out berry picking, putting up food for the winter, and also for the parties that go on in the winter time.

—Seitaan Ed Kunz, L'uknax.ádi

I can't emphasize enough the importance of traditional species to traditional life styles to traditional knowledge. I hope it is clear how these things are all interconnected and interdependent upon one another. Traditional knowledge is born from a relationship with the environment and the practice of cultural activities. The sad thing is that it is all endangered. The knowledge is not getting passed on because the traditions are not being practiced.

And so, our knowledge of the ecology and the environment and our application was, and I'm using this term very carefully, was, the way our people lived. This kind of thing I'm sharing with you right now is hardly part of the educational system that our young people have been growing up in. I grew up in the kind of thing that I'm talking to you about. I was told about to be just appreciative of the water. Just to be able to go to a clean cold little stream. Taking your hands and dipping out the cold water and taking a sip and being able to say gunalchéesh áa, gunalchéesh áa. Thank you. Thank you. Because life is in the water.

—Kingeisti David Katzeek, Shangukeidí

The generations are not being educated. What greater symbol of a culture in distress exists? Many parts of the traditional Tlingit way of life have been transformed by the culture that came to dominate. As European-American society has imposed new systems, the Tlingit lifestyle has deteriorated. Although much work is being done to reestablish the cultural activities, this process is challenged by the current systems in place. So much of the current American rules of engagement stand in opposition to traditional Tlingit society. The methods of interaction have changed in ways that may not be reclaimable. This strikes the core of Tlingit values.

"t belonged to the whole clan. Then, the people can say this is Wooshkeetaan Aaní, this is Eagle land, we can do that. But there were no signs saying keep out. Everybody just knew what belonged to who, what clan. And they always honored that clan. Not one of us can stand up and say this is my property. Although I can now because I put 20,000 bucks on this thing. I had to – conform. And that's what a lot of us American Indians have done; they conformed so we can have some semblance of freedom. You know what freedom is? To breathe. You know, to live? Not too many people know.

—Kaayistaan Marie Olson, Wooshkeetaan

The lost traditions are hard to swallow. The Tlingit way of life seems to be so well balanced compared to how we are living today. One may beg the question, how can the situation be improved? We are aware of what is being lost, how can we fix it before it's gone? These are hard questions to answer. This is a very bothersome subject, so I won't push too much harder on about it here. But before we move on to how to fix the problems, we need to understand why they happened and what is keeping them the way they are. These problems are not quick fixes with the stroke of a government pen. To devise solutions, let us look at a few of the challenges that we are up against.

Challenges to Adaptation

Environmental change is disorienting. To a people who are so deeply connected to a particular environment, the loss of reliability is hard to deal with. Combined with an overall decrease in actually being out there and living in it, the changes seem to be amplified by the lack of time that they have to be able to adapt to the changes. A major challenge then, is the problem of understanding what is going on and how to react to it.

The forest service would like us to talk about this stuff but it's really hard to talk about it when people are in shock over it. You know they're still trying to understand it themselves.

—Koonesh Eric Morrison, L'uknax.ádi

One of the things that I'm worried about, and people know I worry, is that we still need to get to the root causes of why we're here. Why is it, we have such stormy weathers in places we never had before? Why is that the islands are sinking? Why is the glaciers melting? And how much time do we have, when that water starts coming up like it did in the Pacific? And there are no more people living in those little islands. And I pray every morning and I think those people that have lost their homes in the Pacific; that will spread to where we are already losing our homes. In Shishmaref and those villages their homes are lost. It's really important to native people. We are so connected to the land. People criticized the people of Shishmaref because they didn't move. They say, 'you knew it was coming, why didn't you move?' That's because we are so connected to the land. That's why they didn't move.

—Chewshaa Elaine Abraham, K'ineixkwaan

We see that this traditional connection with the world still exists in the modern people. Contemporary distractions of society may be reducing the human being's ability to understand this connection, but it will always be there. The daunting thing is that the distractions are getting more intense, more powerful, more effective. Identifying the value of life beyond the distractions, and helping others to identify this value, is a serious challenge to overcome.

I grew up on Taku River when there was no radio. There was no such thing as cell phones or television or anything like that, everything was candlelight and wood stoves. When you're not cluttering your mind up with TV or cell phones and all that stuff you really become aware of everything that's around you. Even though you don't realize it, it grows on you until you finally find out what's going on... my biggest concern for our people is to get them out of the house and into the country that they were born, get to know it. Take care of it.

—JO-OUAACK John Morris, Yanyeidí

To acknowledge it, to recognize it, to become responsible for it, will help us to interact appropriately with the environment and the ecology. Because if you don't know what the whole ecological type of thing is doing, you're going to totally destroy it. Maybe not because you don't care about it, but because you lack knowledge about how everything is intricately woven together. And that's what's been breaking down. That is what's being torn apart.

—Kingeisti David Katzeek, Shangukeidí

These Tlingit elders have recognized a problem that seems logical, yet has so much resistance to being solved. The less time spent focusing on something, the less it is understood, and the less it can be interacted with appropriately. However the problem is compounded to the 7 billionth degree because this connection is a human need. Every person on the planet is losing this environmental connection in some way, whether they are conscious of it or not. Some are trying to keep it, even strengthen it, but on the large scale, it is being lost to powerful forces. These forces are larger than the power of individuals, but at least they are still human forces.

Sometimes I wonder why people do things but you know, they call it what, progress? ...Those dingalings are now sitting up in the state legislature! They're making rules! ...How are you going to change their thinking when they think they can sell the air that you breathe? They put a monetary value on that! We couldn't make any decisions. The fishermen, they did get angry when they built fish traps at the mouths of the creeks in Southeast. So that they could, the big fishing companies – cannery companies, could make money off the fish. And that was our food! They're taking it out of our mouths! So how are you going to change that?

—Kaayistaan Marie Olson, Wooshkeetaan

It seems that the system in place is not fit for the traditional Tlingit culture. There are values of the culture that are simply contradicted by the current way of doing things. Should we drop everything now and pick up where the Tlingits left off pre-contact? I think that this option is neither possible nor preferable. A complete and sudden overhaul of society would leave most people even more lost than they are now. But the systems are gradually being changed. Positive action is occurring in Southeast Alaska and many Tlingit people are leading the charge.

Adaptation Techniques

> Everything that you do in working within the ecology, within our environment has to be done together. So Austin talked about the importance of being able to begin to return back to the ways our people interacted with the earth in the past... Austin Hammond, has gone so many times, to the Fish and Game, to the Fish and Wildlife Service, even to Washington DC. To express his concern about the environment and the way in which our nation has been handling our earth and not caring about what it does.

—Kingeistí David Katzeek, Shangukeidí

As Tlingit institutions increasingly improve their abilities to conform to American governmental systems, the tribal organizations are able to find more ways to influence society. Traditional cultures are gaining more respect as the knowledge systems are proving to have valuable impacts on the societal decision making process (Cruikshank, 2005). Through my experiences with formal Tlingit education I've seen that this unprecedented level of acceptance is instilling a new wave of confidence in the native people. While there is still much to be done in this regard, the effects can be seen in the youth and their excitement for the cultural way of life. As discussed previously, traditional food harvesting is key to this way of life. Therefore, one of the most important things that modern Tlingit people can do to adapt to American society is to continue the traditional methods of harvesting foods.

> All our people, this is the way we do things you know, when we put up our food. Just clean it, put it in jars, just add salt, and that's all, cook it. No other additives, no chemicals or anything. And a lot of the food that you buy in stores are like that, they got all kinds of preservatives, chemicals, where our people haven't lived with that. And so there are quite a few of our people that are dying from cancer. Because their heart doesn't have any way to fight these things that are being put into the body. So we try to live like the way our old people did. Putting up our foods the old way. Living on the things that we get.

—Seitaan Ed Kunz, Ĺuknax.ádi

Harvesting wild food defines Tlingit culture (Hope and others, 2000). With a short term thought process, going to the grocery store to buy food might be perceived as "easier". However, this method of capturing energy comes along with some less desirable attachments. Industrial agriculture brings a trade out from a subsistence economy to a cash-based economy. The difference comes in how the members of the civilization spend their time. Is a life spent working for money or working for sustenance? As we have seen, the life spent working for sustenance has some obvious implications for health and well being that can be harder to achieve in a capitalist system. These benefits are achieved in different ways today. However, I believe that the time spent participating in activities that secure these benefits replaces the time that is needed to truly comprehend traditional cultural values. Particularly for the Tlingit youth, these values need to be recognized in order to ensure the survival of their culture. Fortunately, the elders are hard at work in establishing programs that educate the youth in traditional ways.

> We're trying to have kids go out and work with the elders as a communal thing. It used to happen all the time when families would go out in the skiffs camping for four or five days to pick berries and smoke fish and it just isn't happening anymore. It's a combination of economics, and the weather, and people with knowledge just aren't as healthy as they used to be. So now we are having them come in here and telling us what you used to do and what you'd like us to do and we'll go out and try to get it. Whether its medicines, or devils clubs, or different roots, or berries, sea asparagus, that sort of thing. Tell us when it's happening, we'll go out and look, see if we can find it. Tell us where you used to gather it, and we'll go explore those areas. They can't get out there but they have the knowledge so it takes a whole different kind of energy to start developing the knowledge base of these guys. On where and when they used to gather it and to take this other group to go out and get some results for that database. So saying ok, there used to be something here, now there's nothing there.

—Koonesh Eric Morrison, Ĺuknax.ádi

So whenever we do this in the summer time we take the Goldbelt catamaran up. We take the kids up there, and I show them all the area up in here. The Taku Glacier, Norris Glacier, Taku Twins, fish camp, berry picking, sacred sites, burial sites, all that kind of stuff. And this is our third year doing that… try to teach these kids and try to get them interested in going into that field in the Fish and Game and the Forest Service. You know instead of just knowing computer stuff they need to get out in the field and know what's out there. And we don't have, when I work with Fish and Game and the Forest Service, a lot of the employees are like my age. They're saying, 'Hey we need to get the young folks interested in this kind of stuff. I'm going to retire soon and there's no one there to take my place.' We really need to get these kids interested in this field so that there's someone there to take over.

—JO-OUAACK John Morris, Yanyeidí

Hopefully programs like these can obtain more support and become a major foundation for the raising of modern Tlingit youth. This depends upon collaboration between both Tlingit and non-Tlingit organizations. It needs to be prioritized by all relevant stakeholders that are involved in the education process; to deliver a well balanced education that consists of both western and traditional knowledge systems. Currently very little is being done by Southeast Alaskan public schools to address this issue. Western knowledge dominates the school environment, leaving little structured time to learn from the elders and cultural knowledge bearers. My prescription for this ailment is a heavy dose of cultural education. Tlingit youth need to be allowed more time with their people in their places, and a greater emphasis needs to be made on learning traditional knowledge. Perhaps the most important form of traditional knowledge to be passed on is the oral narrative story, this method of sharing the cultural values and lessons that have accumulated after centuries of learning and living on the land. These "HiStories" as I have labeled them, are particularly potent sources of information for describing human interactions with environmental changes.

HiStories

Devils Thumb. Taalkunaxk'u Shaa – "The Mountain That Never Flooded"… I have several migration stories. One of them is about The Flood and I know some about the rising flood waters. Like when the Dakl'aweidí left Angoon, there was a woman that didn't want to leave Hood Bay. Yíshx was her name. She went back after her beauty box they said and she got caught in the waters and turned to stone. You can still see her figure in the bay. When the weather's going to change they say the basket she's holding moves. So when they came back years later, maybe centuries perhaps who knows. They knew the areas by landmarks somehow and they worked their way back to Hood Bay from the Stikine River. So they had the story of before the flood, after the flood, and getting caught behind the ice. Which I assume was the last Ice Age.

—Gooch shaayi Harold Jacobs, Yanyeidí

A great flood occurred on this earth. We had to flee our communities. All the communities that you know about right now, people had to flee and go to the mountaintops to survive. We call it The Great Flood… it was called Aangalatkoo- when the earth began to bring up all the water within it, it came from Aangalatkoo, water was coming through the earth upwards as opposed to necessarily just coming down.

—Kingeisti David Katzeek, Shangukeidí

HiStories are likely to be the richest source of traditional knowledge available. However, the availability factor presents difficulty. As HiStories are a form of at.óow, they are the property of the clan that tells the story. Therefore only members of the clan are allowed to share this information. As they are such potent sources of knowledge, they are protected very carefully. Given the recent history of disrespect and ridicule of this type of knowledge, it is not surprising that many culture bearers would be reluctant to share it with uncertain ears.

A lot of times what happens is religious institutions, good ones, I'm not talking against them. Because of their lack of knowledge, have rendered an opinion, that these people are evil, they're paganistic, they're not intelligent, they don't know. Not all religious institutions. But that's almost been the general attitude.

—Kingeisti David Katzeek, Shangukeidí

The Tlingit people have a right to keep their most precious knowledge to themselves. With this difficult history in mind, the researcher of traditional knowledge should be careful about how the topic of HiStories, and all forms of traditional knowledge for that matter, is approached. Particular attention should be given to the types of questions that are asked. In my research, I had mixed success with attempting to document HiStories. Some folks were willing to share short summaries, some could not remember any at the moment, and others either did not understand my request or avoided it all together. I respected each decision and did not let it affect my gratitude for the information that was delivered. I was able to gather a wealth of knowledge about The Great Flood, and I am truly appreciative of that.

My advice for future researchers is to take into account your personal identity and how history precedes your reputation or image. Be aware of your self in interviewing elders. If you are of Tlingit origins, I recommend actively pursuing HiStories and learning as many as you can. The more those stories are shared the better chance they will have of surviving the challenges of time. Be a cultural knowledge bearer and use that knowledge to help your community. For those that are not of Tlingit origins, I recommend taking the backseat approach. Do not seek out HiStories, they do not belong to you. Respect those that have the knowledge and respect their ancestors. However, if you happen to be fortunate enough to hear a fully told HiStory, appreciate it and cherish it. Soak it in and be grateful that you have been in the right place at the right time. Although the research may be directed toward an academic purpose, requesting this type of knowledge cannot be approached from an objective stance. Who you are determines what you can learn. This is perhaps the greatest value of traditional knowledge; the knowledge cannot stand on its own. It depends on living people and personal interactions.

Recommendations

In my work with the Central Council of Tlingit and Haida Indian Tribal Association (CCTHITA) and the University of Alaska Southeast (UAS) on this project, I have learned a great deal about this interface between traditional and western knowledge systems. My lessons came through a decent amount of struggle in trying to figure out how I would comprehend and present these two different ways of knowing in a balanced way. Therefore I feel that it is necessary to provide the readers of this experiential report with some recommendations for how to conduct further research and how to benefit from collaboration and sharing of knowledge.

My first recommendation for the CCTHITA and UAS is to continue seeking outlets for communication. Both institutions could benefit greatly from the resources and information that the other contains. When devising projects, I believe it would be beneficial to always consider how both the central council and the university could contribute to the project. Can the university provide student researchers or literary resources? Does the central council have connections with someone who may know more about this topic? How can an alternative perspective strengthen the assurance of success with this project? These are questions that should be asked with every project concerning local matters.

Different institutions should not only be addressed in the planning and gathering process of projects, but also in the results and application phases. Lessons learned, results of research, next steps, people and places affected, and the like, should be shared at the end of projects. While it seems obvious that local parties should be informed of knowledge that has been gained in their places, it does not always happen (Huntington and others, 2004; AFE Roundtable). This has become such a serious problem that the Canadian government has actually passed laws requiring arctic scientists to communicate with First Nations groups regarding local research projects (AFE Roundtable). While the effectiveness of this measure is still uncertain, it is an action that should begin to be pursued in Alaskan law as well. Scientific findings can be used to inform cultural activities just as developments in Tlingit knowledge can be used to inform solutions to academic problems. As both learning systems grow, the amount of ways that each can benefit the other expands exponentially. As Huntington and others most eloquently put it:

"The benefits of combining TEK with conventional science to give new insights into environmental change can be complemented by the creation of collaborative partnerships, which in addition to producing more and better observations, may also lead to new understandings at the personal level."

—Huntington and others, 2004.

My next recommendation is to follow this advice and create collaborative partnerships. There is a lot of room for students to play a role in the communication process. I encourage students of all ages to continue asking questions about how traditional and academic knowledge systems can benefit one another. Students should be the fundamental unit of information sharing. University students should spend time exploring traditional perspectives, younger Tlingit students should spend time exploring academic perspectives, and vice versa. Ultimately, the best way to accomplish successful information sharing will be to have students that understand both languages involved in collaborative programs. A goal might be to devise a program for every Tlingit community that would be responsible for facilitating the exchange of relevant information. The purpose of the program would be to acquire information that has been produced either traditionally or academically about the local place, and make it available to relevant parties and accessibly consumable. An online database with a well-planned directory could be an ideal medium for this network. It could be student powered and institutionally managed. The Exchange for Local Observations and Knowledge of the Arctic is a great example of this concept. Visit http://eloka-arctic.org/ for more information.

In order to train students to be able to effectively contribute to collaborative projects, the education methods need to be improved. Both CCTHITA and UAS can play powerful roles in this. The need for place-based education is paramount to helping students understand the values of traditional and academic collaboration (Orr, 1995). All scales of the public education system, from preschool to university, need more of a place-based focus. Educators should concentrate on adapting lessons toward local applications. Where the formal education methods are coming short, external institutions should pick up the slack. Extra-curricular programs should focus on connecting the lessons being taught in schools with local applications. Textbook concepts on biology should be applied to local wildlife. Mathematics should be taught with native craftwork. Anthropology lessons should be supplemented by interviews with local historians. The stronger the local knowledge base, the more effective students will be in identifying areas that need improvement and devising solutions to conflicts such as environmental change.

Finally, I hope that readers of this report will use the information presented to do further research. In my work on this project, I encountered an overwhelming amount of sources to guide my research on climate change in Tlingit Aaní. Nearly every person I talked to gave me two other names of people that I needed to talk to. Every paper or book I read referenced countless other published sources. I compiled this information into a Resources for Tlingit Traditional Climate Knowledge[1] document. However, much more organization could be done to make this information more accessible. The Interviews Directory shows a glossary of the interviews with my categories for climate change topics, but the information in the interviews extends well beyond just my categories. I recommend that the reader find something that interests them, either through this report, the interview directory, or the interview transcriptions, and then listen to the audio of the interview. The audio files contain the truest record of what was said, and the hasty written version cannot do it justice. If further intrigued, the student should pursue further conversations with the interviewee, or pursue the sources recommended by said person. Equally important is to ensure that the knowledge bearer benefits from the research. Make it clear that their time will be repaid somehow.

Additionally, keep in mind that one of the flaws of this report is the portrayal of these knowledge systems in an unambiguous light. In order to review these concepts in a clear written format, I have been forced to present traditional and academic knowledge as two separate forces. Such are the limits of the written word. One may have gotten the impression that these ways of learning oppose one another, as they have been labeled with individual definitions. I believe that it is helpful to think of these concepts as different methods of reaching the same goal, the pursuit of understanding our perception of life. If one approaches this type of research from a binary "the one or the other" perspective, I believe he or she will have less success than if one goes in with the understanding that all parties involved have the same goal. Also, be aware that much information hides between the cracks of what is said. While the original intention of this report was documenting climate change, it is apparent that broader ranges of topics were accessed. Do not narrow your focus to one concept. Remain open to all that is being said to you and appreciate the bigger picture messages.

Literally thousands of new research projects could be initiated from this project. Mapping the extent of the Great Flood, developing curricula for teaching traditional knowledge in each level of education, modeling historical species populations, I could go on for pages listing the new questions that I have after doing this research. The most important thing that the reader of this report can do is to develop curiosity about what has been presented. Continue asking questions and encouraging other learners to ask questions about the role of traditional knowledge in modern society. For some final inspiration, I'll end this report with a few of the gems from my interviews. Thanks for reading.

---

[1]These additional resources are available upon request.

Wisdom

Resources are here if we learn how to take care of the resources. How much we can afford to use with money or trade or our own age and the cycle of our environment. That's one thing that my husband keeps reminding me, nobody has a record of when these things started to change. So maybe this is part of it, the record being the cycle that our environment goes through to clean itself up… Those cycles are something that we don't know about. We hardly know our own life cycle because we can't predict when its over, and that way of life too is affected on how we think about our environment. I gave a talk at the Juneau Arts & Culture Center about weaving. And sometimes you're thinking about trying to explain to people that are not weavers but have an interest in the cultural way of life. So I started talking about when I am teaching weaving, I speak to the weavers about being mindful of what they're weaving; how long will it live, beyond you? Cedar bark hat maybe 500 years past your time. The ropes will live thousands of years past your time. So how are you today weaving? How is your mind going around it, wrapping around it? How are we trying to make something that with that in mind? You have the ceremonies they're considering, the transfer of the item, the way of life. The way it's going to be used in ceremonies, the songs that you can make so that it's more useful, more pleasing and have more meaning. So all of these things that we are doing is not just for one time. Like the cycle of our earth, we're going through the cycle of our cultural ways. So that prediction of things is hard to come by. These things will last that long if their taken care of. So like our earth if we don't take care of it, it's going to lose its power to recreate, just like us. Even in women, can't have babies after a certain age, and that's how it is with earth. You can't have constant re-growth when there's constant pounding down, you know, a road being built. So all of that in the cycle of life so to speak, involving earth and us is pretty tenuous. The way we see today just today or tomorrow instead of looking over five hundred years or a thousand or ten thousand years is completely different than what we do as a regular human being on this earth."

—Kasts Saa Waa / Stalth Kaa Waas Della Cheney, Kaach.ádi / Tiits.git.aa.nee

How rich is this community right now from all the gold that was taken out of Mt. Juneau? How wealthy are we? Where's the salmon? Where's the herring? Where's the sea mammals? Where's the gold? They're all gone. They're all gone. Almost all gone. You can't go out there and go catch a fish out in front of the hangar. Or under the bridge. Or outside the boat harbor. They can do that over in Angoon, they can do that over in Huna, they can do that in Sitka. But we can't do it here. But we invested in gold. And so, I'm not against gold. There's a way in which we can do that and do it in a real good balance, and it can be proven to be in good balance. And people aren't cutting corners just to make a dollar and then at the expense of the earth. It's not at the expense of the earth. It's at the expense of mankind.

At the very beginning, I told you

He said because everything has a spirit, and we also have a spirit

He said the spirit world has only one language. Only one language. And we can understand each other in interacting with respect to the environment. He said

If you want to know about the fish, go to the fish

Ask the fish

He said when you ask the fish, the fish will tell you about itself. This is true scientific endeavor. What does the biologist do? The biologist will take a salmon or whatever they're looking at, and whatever they're looking at will talk to the scientist about what the conditions are with respect to that particular study that they're doing. And whatever it may be in salmon, in trees, in water, whatever. It talks to us.

And this is the thing that is hard for western civilization to understand. We say the water talks to us. We say the fish talks to us. We say the tree talks to us. We say the environment the way the weather conditions are is talking to us. Do you hear it?

Do you hear what the earth and the environment is saying to us?

Halt. Stop.

What is being said about this earth? Stop. Listen. Pay attention.

Think about it. Meditate on it. Mull it over in your heart and in your mind. Look at it from every angle. And it will tell you.

Human beings are intelligent. When I say 'ya xoodzi gei Tlingit' most people think I'm talking about a Tlingit person that speaks my language. When we say Tlingit we're saying human beings. But we really emphasize that and the reason why we emphasize being a human being is not a humanistic type of philosophy. Religious worlds want to categorize a whole lot of things. Being able to recognize and accept the intelligence that we have been gifted with regardless of what language you speak, is actually being humble. Because the word I said was gifted. Wasn't something that we did to earn it. But every human being has this intelligence. There's not one human being that doesn't have it.

—Kingeisti David Katzeek, Shangukeidí

When I'm working with the forest service, they say what we really want you to identify is sacred sites. Sacred sites? What do you mean by sacred sites? Burial places? Cemeteries? Fish camps? Exactly tell me what you're asking me. Sacred sites? They say 'well, you know.' And we're dealing with that also with Goldbelt and Sealaska identifying old places that were villages and such. And my answer to them was: 'you know when I hit the mouth of the Taku River, all the way to the Canadian border, to me that is a sacred site. I can't pick one spot and tell you that's a sacred site.' To me that whole area; that whole thing. My part of the world. Like when I told you, you can blindfold me and stick me anywhere on the Taku and I'll tell you where I'm at. So that's how I view it.

—JO-OUAACK John Morris, Yanyeidí

Bibliography for Climate Change in Tlingit Aaní

Individuals interviewed for this report are highlighted with bold text.

Abram, D., 1996, The Spell of the Sensuous – Perception and Language in a More-than-human World: Vintage Books, 326 p.

Connor, C., Streveler, G., Post, A., Monteith, D., and Howell, W., 2009, The Neoglacial Landscape and Human History of Glacier Bay, Glacier Bay National Park and Preserve, Southeast Alaska, USA: The Holocene, v. 19, no. 3, p. 381-93.

Cruikshank, J., 2005, Do Glaciers Listen? - Local Knowledge, Colonial Encounters, and Social Imagination: University of Washington Press, 288 p.

Cruikshank, J., 2001, Glaciers and Climate Change - Perspectives from Oral Tradition: Arctic v. 54, no. 4, p. 377-393.

Dauenhauer, N.M, and Dauenhauer, R., eds., 1993, Glacier Bay History, Haa Shuká = Our Ancestors - Tlingit Oral Narratives: University of Washington press, Seattle, Washington 1993, 514p.

Davis, W., 2009, The Wayfinders: Why Ancient Wisdom Matters in the Modern World. Toronto: House of Anansi, 262 p.

Exchange for Local Observations and Knowledge of the Arctic. 20212, Website: accessed April 12, 2012, at http://eloka-arctic.org/.

Gearheard, S., Pocernich, M., Stewart, R., Sanguya, J., and Huntington, H.P., 2009, Linking Inuit Knowledge and Meteorological Station Observations to Understand Changing Wind Patterns at Clyde River, Nunavut: Climatic Change, doi:10.1007/s10584-009-9587-1.

"Gooch shaayi Harold Jacobs, Yanyeidí." Personal interview. 12 Mar. 2012.

Hennon, P., 2008, Yellow-Cedar Decline: Conserving a Climate-Sensitive Tree Species as Alaska Warms: Deal, R.L., ed., Integrated restoration of forested ecosystems to achieve multiresource benefits – Proceedings of the 2007 National Silvicuture Workshop, General Technical Report PNW-GTR-733, Portland, OR, U.S Department of Agriculture, Forest Service, Pacific Northwest Research Station, p. 233-45.

Hope, A., Thornton, T.F., and Thornton, G.E., 2000, Will the Time Ever Come? - A Tlingit Source Book: Alaska Native Knowledge Network, Center for Cross-Cultural Studies, University of Alaska Fairbanks, Fairbanks, AK, 159 p.

Huntington, H., Callaghan, T., Fox, S., and Krupnik, I., 2004, Matching Traditional and Scientific Observations to Detect Environmental Change: Ambio, Special Report 13, p.18-23.

"JO-OUAACK John Morris, Yanyeidí." Personal interview. 26 Feb. 2012.

"Kaayistaan Marie Olson, Wooshkeetaan." Personal interview. 10 Mar. 2012.

"Kasts Saa Waa / Stalth Kaa Waas Della Cheney, Kaach.ádi / Tiits.git.aa.nee." Personal interview. 16 Feb. 2012.

"Kingeisti David Katzeek, Shangukeidí." Personal interview. 10 Mar. 2012.

"Koonesh Eric Morrison, L'uknax.ádi." Personal interview. 16 Feb. 2012.

Morrison, L'uknax.ádi, Koonesh Eric, and Michelle Davis / EPA. "Climate Change: Our Voices, Sharing Ways Forward." Alaska Forum on the Environment. Dena'ina Convention Center, Anchorage, AK. 7 Feb. 2012. Speech.

Nyman, E., and Leer, J., 2003, Gágiwduł.àt: Brought Forth to Reconfirm - The Legacy of a Taku River Tlingit Clan: Yukon Native Language Centre, Whitehorse, Yukon, Canada, 261 p.

Orr, D.W., 1994, Earth in Mind - On Education, Environment, and the Human Prospect: Island Press, Washington, DC, 221 p.

"Richard Carstensen." Personal interview. 10 Jan. 2012.

"Seitaan Ed Kunz, L'uknax.ádi." Personal interview. 26 Feb. 2012.

Thornton, T.F., 2008, Being and Place among the Tlingit: University of Washington, Seattle, 247 p.

Thornton, T.F., ed., 2012, Haa Léelk'w Hás Aaní Saax'ú = Our Grandparents' Names on the Land: Sealaska Heritage Institute, Juneau.

Turner, N.J., and Clifton, H., 2009, "It's so Different Today" - Climate Change and Indigenous Lifeways in British Columbia, Canada: Global Environmental Change, v. 19, no. 2, p. 180-190.

"X'unei Lance Twitchell, L'uknax.ádi." Personal interview.

Publishing support provided by the U.S.
    Geological Survey Publishing Network,
    Tacoma Publishing Service Center

For more information, contact
    Regional Executive, Alaska
    U.S. Geological Survey
    4210 University Drive
    Anchorage, AK 99508

www.ingramcontent.com/pod-product-compliance
Lightning Source LLC
Chambersburg PA
CBHW080639180526
45168CB00008B/3231